ANNALS OF
THE NEW YORK ACADEMY
OF SCIENCES

Volume 322

EDITORIAL STAFF

Executive Editor
BILL BOLAND

Associate Editor
JOYCE M. HITCHCOCK

The New York Academy of Sciences
2 East 63rd Street
New York, New York 10021

THE NEW YORK
SUMMER AEROSOL STUDY, 1976

ANNALS OF THE NEW YORK ACADEMY OF SCIENCES

Volume 322

THE NEW YORK
SUMMER AEROSOL STUDY, 1976

Edited by Theo. J. Kneip and Morton Lippmann

The New York Academy of Sciences
New York, New York
1979

Library of Congress Cataloging in Publication Data

Main entry under title:

The New York summer aerosol study, 1976.

(Annals of the New York Academy of Sciences; v. 322)
"Papers presented at a joint meeting of the Section of Environmental Sciences and the Section of Atmospheric Sciences of the New York Academy of Sciences."
Includes bibliographical references.
1. Aerosols—Congresses.   2. Air—Pollution—United States—Congresses.
I. Kneip, Theodore, Joseph, 1926–   II. Lippmann, Morton.   III. New York Academy of Sciences. Section of Environmental Sciences.   IV. New York Academy of Sciences. Section on Atmospheric Sciences.   V. Series: New York Academy of Sciences. Annals; v. 322.
Q11.N5 vol. 322 [D884.5] 508'.1s [614.7'12'09747]

79–4647

80 002510
PCP
Printed in the United States of America
ISBN 0–89766–012–9

ANNALS OF THE NEW YORK ACADEMY OF SCIENCES

VOLUME 322

May 14, 1979

THE NEW YORK SUMMER AEROSOL STUDY, 1976 *

Editors
THEO. J. KNEIP AND MORTON LIPPMANN

◆

The New York Summer Aerosol Study (NYSAS), 1976. *By* THEO. J. KNEIP, BRIAN P. LEADERER, DAVID M. BERNSTEIN, and GEORGE T. WOLFF 1

New York Summer Aerosol Study: Number Concentration and Size Distribution of Atmospheric Particles. *By* E. O. KNUTSON, D. SINCLAIR, and BRIAN P. LEADERER 11

Size-Mass Distributions of the New York Summer Aerosol. *By* MORTON LIPPMANN, MICHAEL T. KLEINMAN, DAVID M. BERNSTEIN, GEORGE T. WOLFF, and BRIAN P. LEADERER 29

Light-Scattering Measurements of the New York Summer Aerosol. *By* BRIAN P. LEADERER, DAVID ROMANO, and JAN A. J. STOLWIJK 45

Characterization of Aerosols Upwind of New York City: I. Transport. *By* GEORGE T. WOLFF, PAUL J. LIOY, BRIAN P. LEADERER, DAVID M. BERNSTEIN, and MICHAEL T. KLEINMAN 57

Characterization of Aerosols Upwind of New York City: II. Aerosol Composition. *By* PAUL J. LIOY, GEORGE T. WOLFF, KENNETH A. RAHN, DAVID M. BERNSTEIN, and MICHAEL T. KLEINMAN 73

New York Summer Aerosol Study: Trace Element Concentrations as a Function of Particle Size. *By* DAVID M. BERNSTEIN and KENNETH A. RAHN 87

Chemical Composition of Sulfate as a Function of Particle Size in New York Summer Aerosol. *By* ROGER L. TANNER, ROBERT GARBER, WILLIAM MARLOW, BRIAN P. LEADERER, and MARIE ANN LEYKO 99

Inorganic Nitrogen Compounds in New York City Air. *By* MICHAEL T. KLEINMAN, CAROL TOMCZYK, BRIAN P. LEADERER, and ROGER L. TANNER .. 115

The Nature of the Organic Fraction of the New York City Summer Aerosol. *By* JOAN M. DAISEY, MARIE ANN LEYKO, MICHAEL T. KLEINMAN, and EVA HOFFMAN 125

Gaseous and Particulate Halogens in the New York City Atmosphere. *By* KENNETH A. RAHN, RANDOLPH D. BORYS, ERIC L. BUTLER, and ROBERT A. DUCE 143

A Discussion of the New York Summer Aerosol Study, 1976. *By* PAUL J. LIOY, GEORGE T. WOLFF, and BRIAN P. LEADERER 153

* This *Annal* comprises papers presented at a joint meeting of the Section of Environmental Sciences and the Section of Atmospheric Sciences of The New York Academy of Sciences that was held on April 20, 1977. Paul J. Lioy was the presiding Chairman.

Support for The New York Summer Aerosol Study, 1976 was obtained from a number of sources. These are acknowledged in the introductory paper and individually as appropriate in each paper in the volume.

THE NEW YORK SUMMER AEROSOL STUDY (NYSAS), 1976

Theo. J. Kneip,* Brian P. Leaderer,† and
David M. Bernstein *, ‡

*Institute of Environmental Medicine
New York University Medical Center
New York, New York 10016

† Department of Epidemiology and Public Health
John B. Pierce Foundation Laboratory
Yale University School of Medicine
New Haven, Connecticut 06519

George T. Wolff ‡

Interstate Sanitation Commission
New York, New York 10019

INTRODUCTION

Comprehensive characterizations of urban aerosols have been conducted in Southern California (ACHEX) and St. Louis, Missouri (RAPS). The results of these two studies, however, may not be applicable to the New York City area because of several factors:

1. The nation's highest emission density of fossil fuel burning sources (electrical power generation and automobiles) within the city and surrounding areas;
2. High emission density of industrial sources near the city within the general urban area, which include refineries, chemical plants, scrap metals recovery and refining, etc.
3. The potential for transport of additional pollutants from sources in the highly urbanized and industrialized corridor Washington, D.C. to New York City to Boston on southwest winds, which occur frequently during the summer;
4. The potential for additional transport of sulfates and other small particles from the Midwest;[5] and
5. The existence of a summertime photochemical smog problem that is significantly enhanced by long-range transport.[6]

As a result of these factors a group of laboratories combined their resources to characterize the New York Metropolitan Aerosol. An unusual feature is the truly cooperative nature of the study, with each laboratory participating as part

‡ Current Affiliations:

David M. Bernstein
Brookhaven National Laboratory
Upton, New York 11973

George T. Wolff
General Motors Technical Center
Warren, Michigan 48090

1

0077–8923/79/0322–0001 $01.75/0 © 1979, NYAS

of its ongoing aerosol studies. All specific plans and protocols were developed or modified for the study through discussion by representatives of the several laboratories, and data were circulated as tabulations without interpretive material.

Each topic in this series of papers is discussed by a senior author together with co-authors, all of whom have contributed data and interpretations. The interlaboratory data sharing is clearly demonstrated in the authorship of many of the papers.

This paper discusses the operational aspects of this sampling program and presents an historical overview of what was known about New York City's air at the outset of the program.

OBJECTIVES

The objectives of the study were to characterize the aerosol existing in New York City and the background aerosol being transported into New York City. To accomplish this, simultaneous measurements of as many of the physical and chemical properties of the aerosol as possible were made in New York City and upwind at High Point. In some cases, the various methods used also provided an opportunity to compare data obtained by different techniques and instrumentation. While this was of secondary importance, it was recognized that valuable data would be obtained for comparison of a variety of sampling and measuring instruments. With these goals in mind, we sought the participation of laboratories with experts and interests in a number of different complementary facets of aerosol research. The laboratories, instruments, and types of data obtained are given in TABLE 1.

METHODS

The plan adopted for the program included continuous operation of a number of instruments in New York City and at a rural station at High Point, New Jersey, for the month of August 1976, plus alternating periods of sampling for characterization of mass distributions by particle size and for determination of composition data. (Power limitations prevented simultaneous operation of all equipment at both sites. This program is outlined in TABLE 2. Some instruments were installed and successfully operated well before the official starting date of August 1, 1976, and as much as eight weeks of data were obtained for selected methods.

Site Characteristics

Manhattan

The midtown station is on the roof of the New York University Medical Center residence hall, 14 stories high. The building is located on East 30th Street, between First Avenue and the FDR Drive along the East River. It is in the approximate center of the area having the highest reported concentrations of TSP and SO_2.[1]

TABLE 1

PROGRAM PLAN

Laboratory	Instruments	Results Reported
New York University Med. Cen. (Inst. of Environmental Medicine) (NYU)	Hi-Vol—7-day period.	Mass/trace elements, NO_3^-, $SO_4^=$
	Cyclone particle size classifier—7-day period.	Mass/particle size, Trace elements, NO_3^-, $SO_4^=$
	Hi-Vol—24-hour period.	Organics (with URI)
	GCA-β-Mass	Mass (?)
	Thermo Systems-Piezo Elec.	Mass (?)
	Integrating NO_2 Sampler	NO_2
	Long Path Lasers (2)	Optical attenuation
Yale University (John B. Pierce Foundation) (Yale)	Nephelometer	b-scat.
	Royco	Particle size
	CNC	Particle size
	Andersen (Hi-Vol Cascade Impactor)	Mass/particle size, $SO_4^=$, NO_3^-
	BGI (Hi-Vol Cascade Impactor)	Mass/particle size
	24-hr. Hi Vol	Sulfate (total), NO_3^-
	Meteorology Station	Wind speed, direction, ambient temperature and dew point
Health & Safety Laboratory-ERDA (HASL)	Diffusion battery-CNC	Count/particle size
	EAA	Count/particle size

TABLE 1—*Continued*

Laboratory	Instruments	Results Reported
Brookhaven National Laboratory-ERDA (BNL)	Diffusion Battery—Filters	NH_4^+, NO_3^-, H^+, Sulfur species/particle size
University of Rhode Island (URI)	24-hr. Hi Vol—Quartz filters Andersen (1 cfm)	NH_4^+, NO_3^-, Sulfur species Mass/Particle size, Elemental analysis
	Sierra (Hi-Vol Cascade Impactor)	"
	Halogen sampler	Halogens: Particulates-Vapor/Organic-Inorganic
Interstate Sanitation Commission (ISC)	24-hr. Hi Vol	Organic carbon (with NYU)
	Meteorologic Measurements	Meteorologic analyses
	Royco	Count/Particle size
	Nephelometer	B-scat
	Ozone monitor	O_3
(Data from High Point, N.J. Station)	24-hr Hi Vol (to URI)	To URI and NYU Elemental analysis
Brigham Young University (BYU)		Redox and acid-base analyses of filter samples. $SO_3^=$, $SO_4^=$, NO_2^-, etc.

TABLE 2

EXPERIMENTAL SCHEDULE

TOTAL SUSPENDED PARTICULATE AND MASS/VOLUME-SIZE DISTRIBUTION
OF THE PARTICULATES

(Experiment Duration: Aug. 2–Aug. 9, 1976; Repeated: Aug. 16–Aug. 23, 1976)

Instruments Involved	Laboratory	Operation
Constant Flow Hi-vol (7 day)	NYU	Continuous
Cyclone size class. sampler	NYU	Continuous
Andersen (Hi-vol)	Yale	Continuous
Hi-vol (24-hr)	Yale	Continuous
BGI	Yale	Continuous
Sierra	RI	Continuous
Andersen (1 cfm)	RI	Continuous
Halogens	RI	Continuous
Diff. Batt.	BNL	
Laser Neph.	Yale	Continuous
Charlson Neph.	Yale	Continuous
Royco	Yale	Continuous
CNC	Yale	Continuous
EAA	HASL	6 min/hr
D.B.–CNC	HASL	6 min/hr
β-Mass	NYU-GCA	
Piezo Electric	NYU-Thermo Systems	10 min/hr
Hi-vol (24-hr) Quartz Filter	BNL	6 min/hr
Hi-vol (24-hr) Glass Fiber Filter	Yale	Continuous

COMPOSITION: TOTAL AND SIZE DISTRIBUTION

(Experiment Duration: Aug. 9–Aug. 16, 1976; Aug. 23–Aug. 30, 1976)

Instruments Involved	Laboratory	Filter Media
Cyclone size class. sampler	NYU	Delbag
Lundgren	NYU	Greased Mylar & G.F.
Constant flow hi-vol: AAS metals	NYU	Glass fiber
Andersen (large)	Yale	Glass fiber
Andersen (small)	RI	Polyethelene & Delbag
Sierra	RI	Whatman and Delbag
Halogen	RI	Nuclepore
Diff-Batt	BNL	Quartz
Hi-vol (24-hr) $SO_4^=$ & Sulfur	BNL	4-inch Quartz
Hi-vol (24-hr) $SO_4^=$	Yale, NYU, BYU	Glass fiber
Hi-vol (24-hr) organics	NYU-ISC	Glass fiber
Hi-vol (24-hr) NAA metals	RI-ISC	Delbag
NO_2 & SO_2 integrating	NYU	—

High Point

The High Point station is approximately 88 km west-northwest of New York City and almost equally distant from Scranton and Allentown-Bethlehem. It is situated on a ridge at an altitude of over 480 meters with valleys both east and west some 330 to 360 meters lower. There are no major local sources near this site and the population density in the area is less than 40/km².

Historical Overview

New York City is similar in some respects to other large metropolitan clusters in the northeastern United States. Surface wind velocities in the summer are typically lower than in the winter, 12 km/hour versus 17 km/hour, and the wind directions are highly variable. Frequency diagrams of wind directions for New York generally show winds from all directions in any period chosen, although summer winds are predominantly from the west-southwest, southwest, south and southeast, while more northerly winds may predominate in the winter quarter.

Several continental air masses originated in Canada, passed through the Midwest, and arrived in the New York area on a weekly time cycle during this study. Weather systems (including hurricanes) occasionally move northward along the East Coast and strongly influence the weather in the New York area. The one such event in August 1976 was the hurricane that occurred on the 9th to the 11th and clearly affected air contaminant concentrations on those days.

While differing source terms within New York City cannot be accounted for without a network of sampling sites, major differences in absolute concentrations or in sources in a given area have been defined in New York with as few as three stations.[2-4] In these studies, weekly data for sites in Manhattan, the Bronx, and Queens demonstrated that a single, well-chosen station in the city could be used to establish trends and general conclusions that would be valid for most of the metropolitan area.

Data for TSP and lead are shown in Figures 1 and 2. In addition to demonstrating that the overall interrelationships are common to all sites, it is clearly seen that regulatory steps taken since 1966 have resulted in reductions of the airborne concentrations of these three materials.

An historical data base of trace metal, anionic species concentration, and meteorologic information as well as a history of regulatory actions have been established for the New York University Medical Center site over the past 10 years. In addition, techniques have been developed for apportioning about 70% of the total suspended particulate matter among both natural and anthropogenic sources.[4]

Most of the man-made organic pollutants in New York City's air are believed to result from the combustion of fossil fuels. The largest local sources are automobiles and oil burners. Natural gas combustion generates relatively little organic aerosol, and there is almost no coal burned in the city for either heat or power. The amounts of pollution released by commercial activities and incineration in the city are probably small compared to those for the former sources, and no major industrial sources exist within New York itself.

As the specific characteristics of New York City are not unique, the results of investigations performed there should be broadly applicable to most cities in the Northeast. Most of the cities in this region depend upon oil burning for power and heat and are influenced by long-range transport. The differences between New York City and the others are differences of number rather than of kind. While there is no location in the northeast U.S. whose air quality is not affected by long-range transport, High Point is a good upwind background site because of its location, and the lack of significant local sources or major sources in the vicinity.

The papers that follow detail the physical and chemical properties of the

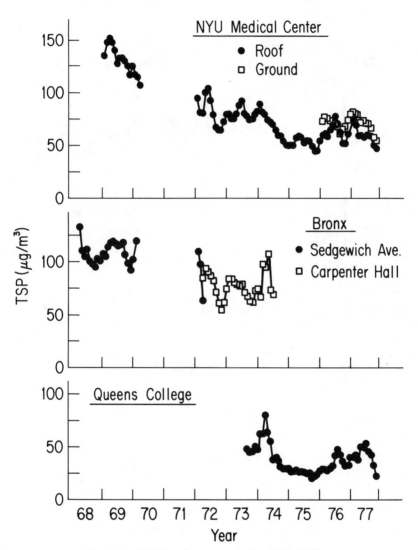

FIGURE 1. TSP Levels at several sites 1968–1977.

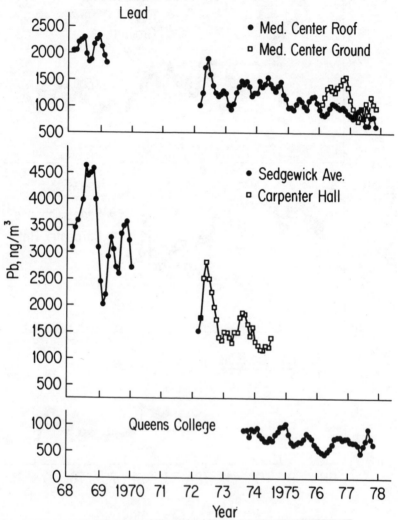

FIGURE 2. Lead levels at several sites 1968–1977.

aerosol, and relate these to the sources and meteorologic factors for the New York area. The comprehensive and unique nature of the results obtained is a direct result of interlaboratory cooperation in all phases of the study from planning through data interpretation.

ACKNOWLEDGMENTS

The efforts of a very large number of people have contributed to the organization and performance of this program. Individuals and sponsors will be acknowledged where appropriate in the following papers.

Mr. Robert Mallon is gratefully acknowledged for his responsible operation of the many systems at the main site at the New York University Medical Center.

Support for the operation of the Medical Center site was provided in part by American Petroleum Institute and Electric Power Research Institute, Grant No. 1058–1, and is part of the center program supported by Grant No. ES 00260, from the National Institute of Environmental Health Sciences, and Grant No. CA 13343 from the National Cancer Institute. Dr. Wolff's research at the High Point station was supported by the Interstate Sanitation Commission and support for Dr. Leaderer's research was provided by the Department of Epidemiology and Public Health, Yale University School of Medicine through Grant No. 5-T01-ES-00123–08, the John B. Pierce Foundation Laboratory through Grant No. 5-S07R-R-05692–07 and the New York State Department of Environmental Conservation.

REFERENCES

1. U. S. Environmental Protection Agency, National Air Quality and Emissions Trends Report, 1975. Pub. No. ERA–450/1–76–002. Research Triangle Park, N.C. (Nov., 1976).
2. KNEIP, T. J., M. EISENBUD, C. D. STREHLOW & P. C. FREUDENTHAL. 1970. Airborne Particulates in New York City. J. Air Pollut. Control Assoc. **20:** 144–149.
3. EISENBUD, M. & T. J. KNEIP. 1976. Trace Metals in Urban Aerosols. Final Report to Electric Power Research Institute, October, 1975. NTIS # Pb–248–324.
4. KLEINMAN, M. T. 1977. The Apportionment of Sources of Airborne Particulate Matter. Doctoral Thesis submitted to the New York University Graduate School of Arts and Sciences.
5. HIDY, G. M., E. Y. TONG & P. K. MUELLER. 1976. Design of a Sulfate Regional Experiment. EPRI Report No. EC–125, February 1976—Electric Power Research Institute. Palo Alto, Ca.
6. WOLFF, G. T., P. J. LIOY, G. D. WIGHT, R. E. MEYERS & R. T. CEDERWALL. 1977. An Investigation of Long Range Transport of Ozone Across the Midwestern and North-Eastern U.S. Atmos. Environ. **11:** 797–802.

NEW YORK SUMMER AEROSOL STUDY:
NUMBER CONCENTRATION AND SIZE DISTRIBUTION OF ATMOSPHERIC PARTICLES

E. O. Knutson and D. Sinclair

Environmental Measurements Laboratory
United States Department of Energy
New York, New York 10014

Brian P. Leaderer

Department of Epidemiology and Public Health
John B. Pierce Foundation Laboratory
Yale University School of Medicine
New Haven, Connecticut 06519

INTRODUCTION

In August 1976, the number concentration and size distribution of atmospheric particles in New York City were measured on an hourly basis as part of the New York Summer Aerosol Study (NYSAS). NYSAS was an intensive, multi-laboratory effort to characterize the New York urban aerosol under summer-time conditions. The sampling site, array of instruments and other circumstances surrounding NYSAS are described elsewhere.[1] In this paper, we present measurements of the particle number and size, and show changes in number and size with time of day and meteorologic variables. Relationships to source characteristics are also explored. Other papers in the series will delineate the relationships of the number/size data to other NYSAS data.

APPARATUS

Measurements of the number concentration and particle size were obtained by simultaneous operation of three instruments:

1. The Sinclair diffusion battery and nucleus counter;
2. An electrical aerosol analyzer;
3. An optical single-particle counter.

This combination of instruments covered the particle diameter range 0.0032 to 5.0 μm.

The Sinclair diffusion battery and nucleus counter is an aerosol measurement system made up of a compact diffusion battery,[2] a continuous flow condensation nucleus counter,[3] and peripheral equipment to permit continuous, automatic operation. FIGURE 1 is a schematic of the system, denoted as ATB-CFC.

The compact diffusion battery[2] consists of a series of metal discs 4 cm in diameter called collimated holes structures, each having 14,500 parallel holes of uniform diameter. The present system has 13 outlet taps: one for sampling the inlet aerosol, one for each of the 11 stages of diffusional separation, and one down-stream of the absolute filter that is included for zero-checking the nucleus counter.

11

0077–8923/79/0322–0011 $01.75/0 © 1979, NYAS

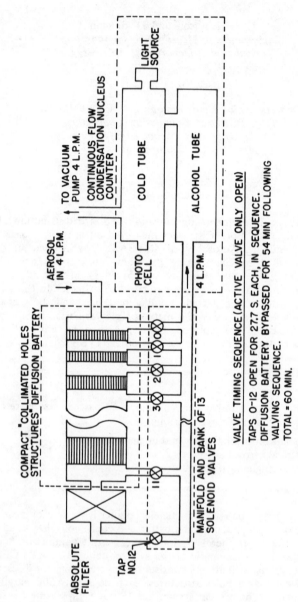

FIGURE 1. Sinclair diffusion battery and nucleus counter (ATB-CFC).

The continuous flow condensation nucleus counter [3] was used to measure the concentration of the aerosol penetrating the various diffusion battery stages. Flowing at a uniform rate, the aerosol sample passes over an ethanol pool, and then through a thermoelectrically cooled condensation tube ($-15°$ C). Ethanol vapor condenses on the aerosol particles forming droplets which attenuate the light beam. The nucleus counter was calibrated for NYSAS operating conditions against a standard Pollak nucleus counter.[4] An ammonium sulfate spray aerosol at concentrations from 300 to 250,000 particles/cm[3], was used for this calibration.

Connection of the diffusion battery to the nucleus counter was made through a bank of 13 solenoid valves, which were operated on an hourly cycle. In each cycle, the 13 valves opened in sequence for 27.7 sec each, starting with the valve numbered zero. The 6-minute valving sequence was followed by a 54-minute idle period during which the diffusion battery was bypassed. The cycles were initiated every hour on the hour.

The electrical aerosol analyzer (EAA), a commercial unit (Thermo-Systems, Inc., St. Paul, MN; Model 3030 Electrical Aerosol Size Analyzer), is described in the literature.[5-8] The EAA was operated in the "single scan" mode, synchronized with the hourly cycle of the diffusion battery.

The optical single-particle counter (OPC) was also a commercial unit (Royco Instruments, Menlo Park, CA, Model 225 with the Model 518 module). The OPC operated on its internal clock, providing a count of particles in five size classes every 10 minutes. The low sampling rate (0.283 ℓpm) was used. Prior use of the Royco 225 for atmospheric sampling has been reported.[9]

The three instruments were housed in a 14th-floor rooftop shed, arranged as shown in FIGURE 2. The air sample was taken in through a vertical tube (3.5 cm in diameter), which projected 0.6 m above the ridge of the shed's roof. The inlet was equipped with a rain hat (skirt diameter = 12.6 cm) and an 18-mesh screen to bar insect entry. Air movement to the inlet was unobstructed in all directions, except for a minor obscuration of ESE winds by an elevator service tower. The total flow in the vertical sampling tube was 59 ℓpm.

To suppress the wide and rapid excursions of concentration characteristics of urban aerosols, a 60-liter holding tank was interposed between the sampling tube and the ATB-CFC and EAA. This increased the average age of the aerosol by 7.5 minutes, equivalent to a 1 km of horizontal travel at the average wind speed. It was subsequently calculated that coagulation in the 7.5 minutes caused a 40% drop in concentration of 0.005-μm-diameter particles and a 15% drop for 0.01-μm-diameter particles at a typical NYC aerosol concentration. We also estimate that wall losses in the holding tank were negligible.

PROCEDURE

Operation of the three instruments commenced on August 13, 1976. The instruments were checked twice daily and minor adjustments were made to keep the operating parameters within their nominal range. Sampling was stopped on September 4, 1976.

The signals from the ATB-CFC and the EAA were recorded on a dual pen strip chart recorder. Each hour's data consisted of 13 voltage levels from the ATB-CFC and 11 from the EAA. The OPC counts were printed every 10 minutes by means of the OPC's printer accessory. The 10-minute count occurring nearest the hour was taken as the hourly reading.

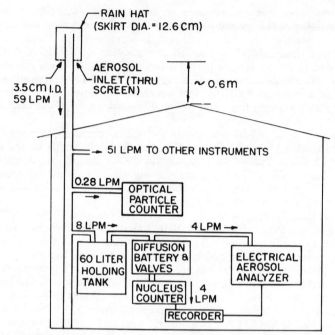

FIGURE 2. Apparatus arrangement for NYSAS number/size measurements.

The data recorded on the strip chart were manually transcribed to computer cards. An empirical calibration equation was used to convert the ATB-CFC voltage readings into aerosol concentrations in particles/cm³. Twomey's non-linear iterative algorithm [10] was then applied to obtain the aerosol concentration in eight size classes. These classes spanned the particle diameter range 0.0032 to 0.32 μm in a geometric progression.

The hourly EAA data were processed according to the instrument operating manual to obtain the aerosol concentration in seven size classes spanning the particle diameter range 0.01 to 0.56 μm. The OPC data gave the aerosol concentration in four size classes (0.5–0.63; 0.63–1.3; 1.3–3.0; 3.0–5.0 μm).

TABLE 1 summarizes the particle size range coverage obtained by use of these three instruments. The overlap in coverage by the ATB-CFC and the EAA provided a useful cross-check on the data in part of the particle size spectrum.

The data from the three aerosol instruments were supplemented by meteorologic data obtained hourly at the sampling site. These data included wind speed, wind direction, ambient temperature, and dew point. The relative humidity was computed from the temperature and dew point.

RESULTS

Number-Weighted Size Distribution

FIGURE 3 shows the median number-weighted aerosol size distribution as determined from the three number/size instruments used in NYSAS. The medians for the ATB-CFC are based on the 230 hours for which ATB-CFC and meteorologic data were simultaneously available. The medians for the EAA and OPC are based on the 410 hours for which EAA, OPC and meteorologic data were simultaneously present. Most data hours in the 230 hour set were represented also in the 410 hour set.

It is seen in FIGURE 3 that the ATB-CFC median results differ from those of the EAA by as much as a factor of two for individual size classes within the range of overlap. We believe that the ATB-CFC data are more accurate below 0.1 μm and that the EAA data are more accurate above 0.1 μm. (This question does not affect the main conclusions of the present paper.)

FIGURE 3 shows that the mode of the median number-weighted size distribution occurs at 0.024 μm. The geometric mean particle diameter computed from the ATB-CFC data had a median value of 0.026 μm.

Volume-Weighted Size Distribution

Volume-weighted size distributions are presented in FIGURE 4 for two subsets of the data. One curve is for a 96-hour period from noon on August 25 to noon on August 29, during which a pollution episode occurred in New York City. The second curve pertains to the remaining 314 hours of data.

The volume/size curve for the 314-hour data set is a familiar one for urban aerosols.[11] The mode is at 0.2 μm diameter and the minimum in the range 1.0 to 2.0 μm. There is a suggestion of a second mode beyond 4.0 μm.

The volume/size curve for the pollution episode is quite different for the larger sizes. The region from 1.0 to 2.0 μm is completely filled in, so that the mode is now at 0.7 to 0.8 μm. The particle concentration in the 1–2 μm range is 8 to 10 times greater than that for the remainder of the period. Presumably, there are a minimum and second mode beyond 4.0 μm.

TABLE 1

PARTICLE SIZE CLASSES FOR THE NYSAS DATA

Size Class Number	Particle Diameter at Class Geometric Midpoint (μm)		
	ATB-CFC *	EAA †	OPC ‡
1	0.0042		
2	0.0075		
3	0.0133	0.0133	
4	0.0237	0.0237	
5	0.0422	0.0422	
6	0.075	0.075	
7	0.133	0.133	
8	0.237	0.237	
9		0.422	
10			0.56
11			0.90
12			1.98
13			3.88

* Sinclair Diffusion Battery and Nucleus Counter—data hours 233.
† Electrical Aerosol Analyzer—data hours 723.
‡ Optical Particle Counter—data hours 432.

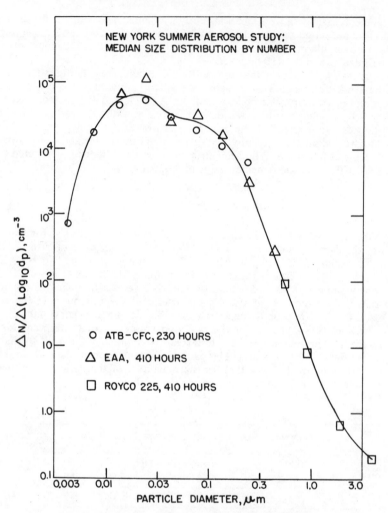

FIGURE 3. Median number-weighted size distribution.

The EAA and OPC data for the pollution episode showed a sharp mismatch, in contrast to the remainder of the period for which the average data from the two instruments faired together smoothly. This contrast is evidence of a difference in either optical or electrical properties of the aerosol for the two periods. The true curve for the pollution episode is probably best indicated by the faired-in curve shown in FIGURE 4.

Correlations to Meteorologic Factors

The data were screened for relationships among variables by computing correlation coefficients. These coefficients provide an overview of the data and serve as a guide for more detailed examination by means of plots. TABLE 2 gives the key correlation coefficients.

The values under time of day and wind direction in TABLE 2 require explanation. Since time of day is a cyclic variable, it was represented in these calculations by the Fourier terms sin T, cos T, sin 2T, cos 2T, sin 3T, cos 3T, sin 4T and cos 4T, where T is the time of day in radians. For each particle size class, the value listed in TABLE 2 is the largest (in absolute value) of the correlations between the aerosol concentration and the eight terms sin T, . . . cos 4T. Similarly, the wind direction was represented by eight Fourier terms sin θ, . . ., cos 4θ, and only the largest of the eight correlations is shown in TABLE 2.

TABLE 2 suggests that the full particle size spectrum comprises two (and possibly three) subranges with differing relationships of aerosol concentration with meteorologic variables. A division point at 0.1 μm is indicated by the correlations with time of day and with temperature, dew point, and relative humidity. This classification is supported independently by the ATB-CFC data and the EAA data. A second division point at 0.5 μm is suggested by the fact that the concentration for 0.1- to 0.5-μm-diameter particles correlates better with dew point and temperature than with relative humidity, while the reverse applies for diameters above 0.5 μm. This division point, however, could also be due to the occasional mismatch between the EAA and OPC data.

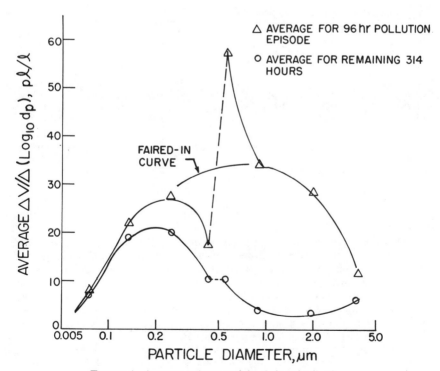

FIGURE 4. Average volume-weighted size distribution.

TABLE 2

CORRELATION COEFFICIENTS RELATING AEROSOL CONCENTRATION
AND METEOROLOGIC FACTORS

Particle Diameter (μm)	Meteorologic Factors					
	Time of Day *	Wind Direction †	Wind Speed	Ambient Temperature	Dew Point	Relative Humidity
ATB-CFC Data						
0.0042	0.23	0.13	−0.01	0.19	0.08	−0.16
0.0075	0.30	0.29	−0.13	0.15	0.13	−0.03
0.0133	0.27	0.41	−0.14	−0.03	−0.06	−0.06
0.024	0.30	0.34	−0.14	0.05	0.01	−0.07
0.042	0.38	0.16	−0.25	0.13	0.10	−0.05
0.075	0.27	0.17	−0.30	0	0	−0.01
0.133	0.25	0.20	−0.31	0.14	0.20	0.06
0.24	0.22	0.19	−0.18	0.33	0.41	0.10
EAA Data						
0.013	0.27	0.41	−0.41	−0.06	0	0.11
0.024	0.46	0.27	−0.20	0.02	0.02	0.01
0.042	0.36	0.21	−0.30	−0.03	0	0.05
0.075	0.25	0.34	−0.38	0.03	0.09	0.10
0.133	0.19	0.32	−0.34	0.33	0.37	0.20
0.24	0.11	0.18	−0.27	0.51	0.58	0.34
0.42	0.11	0.18	−0.23	0.44	0.55	0.38
OPC Data						
0.56	0.11	0.44	−0.30	0.25	0.42	0.45
0.90	0.15	0.44	−0.29	0.17	0.35	0.44
1.98	0.20	0.40	−0.27	0.12	0.30	0.42
3.88	0.15	0.42	−0.22	0.08	0.17	0.23

* The highest correlation among eight sine-cosine terms used to represent diurnal variations.
† As in *, but for wind direction.

Diurnal Variations

Plots of the average concentration as a function of time of day showed that the most pronounced diurnal variation of aerosol concentration was that for 0.024-μm-diameter particles. FIGURE 5 shows this diurnal pattern in terms of the arithmetic average concentration. The pattern is characterized by a sharp drop in concentration between 11:00 P.M. and 5:00 A.M., followed by an equally rapid fivefold increase between 5:00 A.M. and 10:00 A.M. The diurnal pattern for the 96-hour pollution episode and for the balance of the data were identical to the grand average pattern shown in FIGURE 5.

Presumably, diurnal patterns in aerosol concentration are the result of diurnally varying source strengths, dispersion conditions, and humidification growth processes. FIGURE 5 shows the diurnal patterns of two major urban sources of aerosol. The average traffic volume on FDR Drive, 14 floors below the sampling site, was computed from hourly traffic counts made during the experiment period. The curve is similar to one reported for Hudson County, N.J.,[12] and is believed to be representative of traffic in the entire metropolitan area. The curve for electric power production was taken from the literature [12]; data obtained subsequently from Consolidated Edison in New York City showed a very similar pattern.

FIGURE 5 leaves little doubt that 0.024-μm particles in New York City are mainly due to traffic. The rapid fall-off of traffic after midnight is followed with a 1½-hour time lag by a nearly proportional decrease in aerosol concentration. Apparently, dispersion and coagulation are adequate to bring the concentration down after the traffic subsides. (We have estimated mathematically that coagulation alone would produce the rate of decrease shown in FIGURE 5.) Similarly, the increase in traffic volume during the morning rush hour is followed shortly by a nearly proportional increase in aerosol concentration. The proportionality appears to break down somewhat in the afternoon, possibly because of different dispersion conditions.

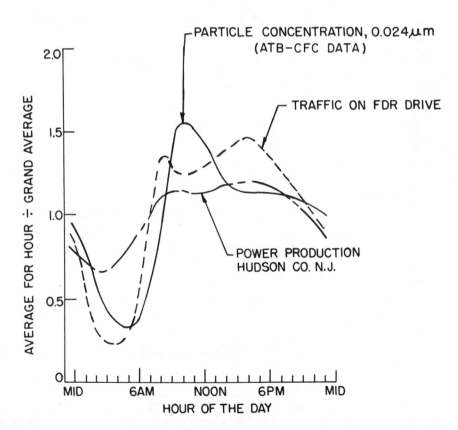

FIGURE 5. Diurnal variations of aerosol concentration, traffic and power production in New York City.

The diurnal patterns for other particle sizes below 0.1 μm in diameter were similar to that for 0.024 μm, although less pronounced. Therefore, we believe that tailpipe emissions are the dominant source of particles below 0.1 μm in New York City. Measurements near freeways [13, 14] have shown that traffic is a rich source of small particles, with relatively few larger than 0.15 μm except under unusual conditions favoring coagulation.

Plots similar to FIGURE 5 showed that particles in the 0.1 to 0.5 μm diameter range had barely discernable diurnal patterns of concentration. Their only feature was a small maximum at 8:00 A.M. This near lack of pattern for New York is in contrast to Los Angeles, where the concentration in this size range increases prominently near noon as a result of photochemical processes.[11] We think that in New York the local contribution to the 0.1- to 0.5-μm range consists of coagulated automobile exhaust particles.

For particles larger than 0.5 μm in diameter, the diurnal pattern of concentration was opposite to that for particles less than 0.1 μm. The pattern for

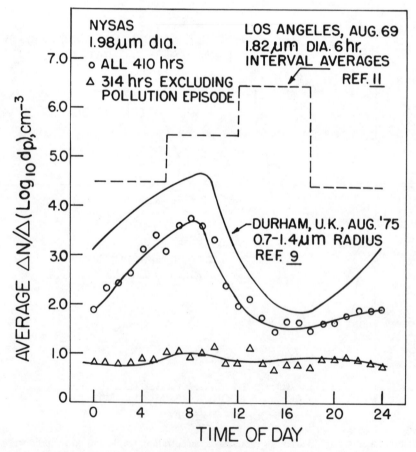

FIGURE 6. Diurnal variations of concentration for large particles.

TABLE 3

VARIATION OF AEROSOL CONCENTRATION WITH WIND DIRECTION *

Particle Diameter (μm)	Median Concentration for Given Direction ÷ Overall Median						
	SE	S	SW	W	NW	N	NE
0.0042	1.51	1.55	0.67	0.22	1.53	0.97	1.56
0.0075	1.78	1.13	0.85	0.23	0.98	2.11	1.28
0.0133	1.71	1.15	1.07	0.32	0.82	2.42	1.43
0.024	1.16	1.18	0.82	0.53	1.04	1.65	1.32
0.042	0.73	1.27	0.89	0.77	0.96	1.57	0.99
0.075	0.87	1.13	0.95	1.01	0.86	1.12	0.91
0.133	0.95	1.03	1.14	1.24	0.88	0.78	0.77
0.24	1.04	0.83	1.06	1.17	1.00	0.67	0.71
0.42	1.25	1.00	1.00	1.00	1.00	1.00	1.00
0.56	2.52 †	4.84 †	4.50 †	2.39 †	0.72	0.64	0.91
0.90	2.38 †	5.31 †	4.90 †	2.53 †	0.64	0.56	0.89
1.98	2.28 †	4.80 †	4.10 †	2.28 †	0.76	0.56	0.86
3.88	1.09 †	1.56 †	1.38 †	1.42 †	1.00	0.69	0.77

* East is omitted because of the low number of data hours for this direction.
† These values are strongly affected by data from the 96-hour pollution episode.

these larger particles, however, was nearly masked by large day-to-day changes. Furthermore, the pattern nearly disappeared when the data from the 96-hour pollution episode were excluded.

FIGURE 6 shows the observed patterns for particles 1.98 μm in diameter along with two curves from the literature. It is seen that our pattern based on all data is very similar to one reported for Durham in the United Kingdom, but different from that reported for Los Angeles. Los Angeles appears to have more afternoon aerosol than New York.

Variations with Wind Direction and Speed

The aerosol concentration varied significantly with wind direction through most of the size spectrum. TABLE 3 shows median values of the concentration.

For particles having diameters less than 0.05 μm, the aerosol concentration showed a minimum for west wind and a maximum for north winds. The minimum was due in part to the fact that west winds occurred most frequently between midnight and 8:00 A.M., when the traffic aerosol source was at minimum strength. The high concentrations found for north winds are probably due to traffic on the FDR Drive, which extends 7 km north-northeastward from the sampling site.

The particle diameter range 0.05 to 0.5 μm showed only moderate variations of concentration with wind direction. The patterns were reversed as compared to those for particles less than 0.05 μm in diameter.

The high median concentrations shown in TABLE 3 for SE, S, SW, and W wind occurred because the wind was in these sectors through most of the 96-hour pollution episode. FIGURE 7 shows this effect for 1.98-μm particles. The medians based on all data indicate a large excess concentration for SE, S,

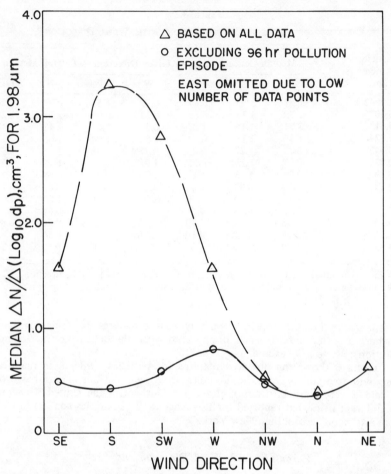

FIGURE 7. Wind direction variations of concentration for 1.98 μm diameter particles.

SW, and W winds. After excluding the 96 hours of pollution-episode data, the median concentration varies more moderately with wind direction, with a maximum at west and a minimum at north. The 96-hour period accounted for approximately ⅕, ½, ½, and ⅓ of the total hours of SE, S, SW, and W winds, respectively.

Relationships between wind speed and aerosol concentration are obscured by the fact that the average wind speed varied with wind direction. (Thus, the patterns in TABLE 3 may result in part from variations in wind speed.) To further explore these interrelationships, partial correlation coefficients between aerosol concentration (at each particle size) and wind speed were computed after removing the effects of time of day and wind direction by means of a multiple regression. These partial correlation coefficients showed that for particles in the 0.024- to 0.24-μm diameter range, the aerosol concentration related significantly to wind speed. Weak correlations were found for both smaller and larger particles. All correlation coefficients were negative.

The relationship between aerosol concentration and mixing height was explored using 7:00 A.M. sounding data from the National Weather Service. Plots were made of the 10:00 A.M. aerosol concentration versus the product of the mixing height and the wind speed through the mixing layer (7:00 A.M. readings). The correlations were nil except in the 0.1- to 0.5-μm diameter range, where they were only moderate.

Variations with Temperature, Dew Point, and Relative Humidity

It was indicated in TABLE 2 that the aerosol concentration for particles smaller than 0.1 μm in diameter had negligible correlation to temperature, dew point, or relative humidity. Above 0.1 μm diameter, correlations developed rapidly with increasing size.

In the particle diameter range 0.1 to 0.5 μm, the aerosol concentration correlated better with dew point and temperature than with relative humidity. The largest correlation found (0.58) was between aerosol concentration at 0.24-μm diameter and dew point. The corresponding partial correlation coefficient after removing time of day and wind direction variations was 0.71. FIGURE 8 shows the median aerosol concentration at 0.24-μm diameter for several dew point classes.

As seen in FIGURE 8, there is a linear relationship between aerosol concentration (at 0.24 μm) and dew point, at least in the dew point range 8 to 20° C. This is not an instrumental artifact, since identical trends were obtained with two independent instruments. Furthermore, it is unlikely that the trend shown is a disguised effect of other meteorologic variables, since the correlation to dew point was higher than to the other variables considered. Deletion of the data from the 96-hour pollution episode had no effect on the trend, as seen by comparison of the 314-hour and the 410-hour data in FIGURE 8.

The curves in FIGURE 8 are shown as turning over at high dew points, but this is uncertain because there are relatively few hours of data with dew point over 26° C.

For particles larger than 0.5 μm in diameter, the correlations of concentration with relative humidity were better than those with temperature or dew point. Most of this correlation, however, was due to the pollution episode, which brought very high concentration of particles larger than 0.5 μm, together with high humidity. When these 96 hours were deleted from the data set, the aerosol concentration showed only moderate variation with relative humidity.

Particle Shape

Eight one-day samples for electron microscopy were taken during the measurement period by means of an electrostatic sampler (Thermo-Systems, Inc., St. Paul, MN; Model 3100 Electrostatic Aerosol Sampler). The samples were collected on carbon foil grids and examined with an electron microscope (JEOL, Medford, MA; Model JEM-100C) in the transmission mode. FIGURE 9 is an electron micrograph of a sample taken on the first day of the four-day pollution episode. Shadowing was not used in these micrographs.

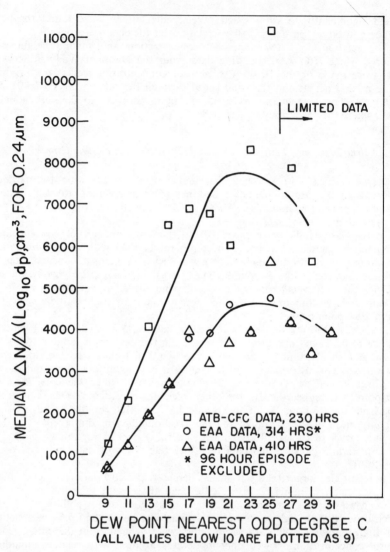

FIGURE 8. Relationship between aerosol concentration and dew point.

FIGURE 9 reveals at least two types of particles larger than 0.1 μm: occasional chain agglomerates of small primary particles and more abundant particles with circular to hexagonal outlines. Only a few particles with diameters below 0.1 μm can be seen in FIGURE 9, because of their very poor contrast. Cubical shapes, probably NaCl crystals, were seen occasionally.

The most striking attribute of the circular-to-hexagonal particles was their volatility. They were easily destroyed in the electron beam. Quite often, particle disappearance was preceded by a "bubbling" effect, indicating a liquid state prior to disappearing.

DISCUSSION

Several of the patterns described in the preceding section are unexpected and cannot be explained on the basis of data presently available. In this section, we identify these patterns and offer hypotheses.

On the average, the concentration of particles above 0.1 μm in diameter showed only moderate variation with time of day, tending to increase overnight to a maximum at about 8:00 A.M. These patterns differ in phase and amplitude

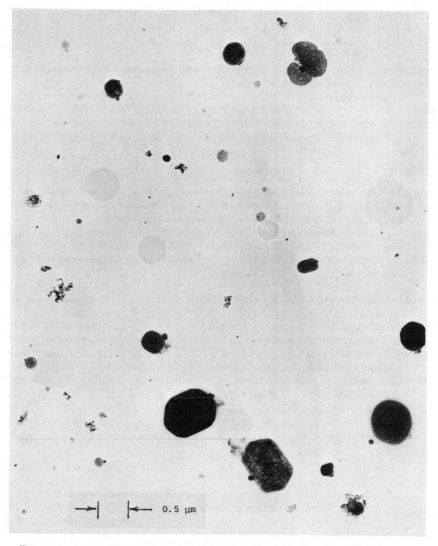

0.5 μm

FIGURE 9. Transmission electron micrograph of New York City aerosol particles.

from the diurnal variations of traffic and electric power production (shown in FIGURE 5) and probably most other local sources. It is possible that diurnal variations in atmospheric dispersing power, humidification growth and other processes may act in concert to cancel or reverse source strength patterns. A simpler hypothesis, however, is that the aerosol we measured was an accumulation of emissions from several urban areas, so that diurnal patterns of source strength "washed out" because of varying transit times to New York City. This presupposes a mean atmospheric residence time of at least a few days for particles above 0.1 μm diameter.

The hypothesis that the aerosol we measured in the 0.1 to 5.0 μm diameter range was regional, rather than New York City urban, provides a simple explanation for the rather mild diurnal patterns we observed in this size range. The same hypothesis would explain the mild variation with wind direction which we observed after excluding data from the pollution episode. This hypothesis should be tested by simultaneous measurements with equivalent apparatus at two locations, urban and nonurban.

A second intriguing feature in our data was the correlation between concentration in the 0.1- to 0.5-μm diameter range and dew point. This suggests a hygroscopicity effect. If this were the case, however, we should expect a stronger correlation to relative humidity than to dew point. Therefore, other possible explanations for the correlation to dew point should be explored. These include:

1. Atmospheric water vapor and particles in the 0.1- to 0.5-μm diameter range may have common geographic source areas, so that air masses with high dew point also tend to have high aerosol concentration.
2. The dew point may be a good index of atmospheric dispersion, so that poor dispersion leads to high aerosol concentration and high dew point.
3. Water vapor may tend to catalyze atmospheric reactions which form particles in the 0.1- to 0.5-μm range.

Our data do not permit us to select between these and other hypotheses concerning the dew point correlation.

A third prominent feature of the NYSAS number/size data was the four-day pollution episode from August 25 to August 29. This period is remembered as hot, humid, and uncomfortable, with S, SW, and W winds. However, analysis of our data excluding the pollution episode shows that neither temperature, humidity, nor wind direction taken individually could explain or predict the high concentrations of 0.5 μm and larger particles during the episode. The mixing height was unusually low only on the first day of the episode. Perhaps the episode can be explained in terms of a combination of factors, or in terms of their persistence.

SUMMARY

The major findings of the number/size measurements made as part of NYSAS are as follows:
1. Most particles below 0.1 μm in diameter in New York City derive mainly from traffic and are probably primary tailpipe particles rather than products of reactions in the atmosphere.

2. The volume-weighted size distribution showed a mode in the 0.1- to 0.5-μm diameter range. In contrast to the Los Angeles aerosol, there was no evidence of a strong daily cycle of photochemical input to this size range.

3. A radical change in the size distribution above 0.1 μm, and especially above 0.5 μm, occurred during a four-day pollution episode. There was evidence of a change of composition as well, based on the different responses of two of our aerosol instruments during the episode.

4. The best single predictor for concentration in the 0.1 to 0.5 μm diameter range was the dew point.

5. Apart from the pollution episode, the concentration of particles above 0.1 μm in diameter showed only moderate (factor of 2 or less) variations with time of day and the meteorologic factors. An exception was the dew point effect mentioned in item 4 above.

ACKNOWLEDGMENTS

The authors wish to acknowledge many helpful discussions with the collaborators in the New York Summer Aerosol Study, and with colleagues at the Environmental Measurements Laboratory and Yale. We especially thank L. Hinchliffe of the Environmental Measurement Laboratory for preparing and interpreting electron micrographs. Partial support for Dr. Leaderer's research was provided by the Department of Epidemiology and Public Health, Yale University School of Medicine through Grant No. 5-T01-ES-00123–08, the John B. Pierce Foundation Laboratory through Grant No. 5-S07-RR-05692–07 and the New York State Department of Environmental Conservation.

REFERENCES

1. KNEIP, T. J., B. P. LEADERER, D. M. BERNSTEIN & G. T. WOLFF. The New York Summer Aerosol Study (NYSAS), 1976. In press. Ann. N.Y. Acad. Sci. This volume.
2. SINCLAIR, D. 1972. A portable diffusion battery. Am. Ind. Hyg. Assoc. J. **33**: 729.
3. SINCLAIR, D. & G. S. HOOPES. 1975. A continuous flow condensation nucleus counter. J. Aerosol Sci. **6**: 1.
4. POLLAK, L. W. & A. L. METNIEKS. 1959. New calibration of photo-electric nucleus counters. Geofis. Pura Appl. **43**: 285.
5. LIU, B. Y. H., K. T. WHITBY & D. Y. H. PUI. 1974. A portable electrical analyzer for size distribution measurement of submicron aerosols. J. Air Pollut. Control Assoc. **24**: 1067.
6. LIU, B. Y. H. & D. Y. H. PUI. 1975. On the performance of the electrical aerosol analyzer. J. Aerosol Sci. **6**: 249.
7. SEM, G. J. 1975. Design and application of an electrical size analyzer for submicron aerosol particles. Instrument Society of America Reprint, AID 75408. Pittsburgh, PA.
8. WHITBY, K. T. 1976. Electrical measurement of aerosols. *In* Fine Particles: Aerosol Generation, Measurement, Sampling and Analysis. B. Y. H. Liu, Ed., Academic Press. New York.
9. JENNINGS, S. G. & R. K. ELLESON. 1976. Aerosol particle size distributions in the 0.25–5.0 μm radius range in Northern England. Atmos. Environ. **11**: 361.
10. TWOMEY, S. 1975. Comparison of constrained linear inversion and an iterative nonlinear algorithm applied to the indirect estimation of particle size distributions. J. Comp. Phys. **18**: 188.

11. WHITBY, K. T., R. B. HUSAR & B. Y. H. LIU. 1972. The aerosol size distribution of Los Angeles smog. J. Colloid Interface Sci. **39:** 177.
12. GRAEDEL, T. E., L. A. FARROW & T. A. WEBER. 1976. Kinetic studies of the photochemistry of the urban troposphere. Atmos. Environ. **10:** 1095.
13. WHITBY, K. T., W. E. CLARK, V. A. MARPLE, G. M. SVERDRUP, G. J. SEM, K. WILLEKE, B. Y. H. LIU & D. Y. H. PUI. 1975. Characterization of California aerosols—I. Size distributions of freeway aerosol. Atmos. Environ. **9:** 463.
14. WILSON, W. E., L. L. SPILLER, T. G. ELLESTAD, P. J. LAMOTHE, T. G. DZUBAY, R. K. STEVENS, E. S. MACIAS, R. A. FLETCHER, J. D. HUSAR, R. B. HUSAR, K. T. WHITBY, D. B. KITTELSON & B. K. CANTRELL. 1977. General Motors sulfate dispersion experiment: summary of EPA measurements. J. Air Pollut. Control Assoc. **27:** 46.

SIZE-MASS DISTRIBUTIONS OF THE
NEW YORK SUMMER AEROSOL *

Morton Lippmann, Michael T. Kleinman,† and
David M. Bernstein †

*Institute of Environmental Medicine
New York University Medical Center
New York, New York 10016*

George T. Wolff †

*Interstate Sanitation Commission
New York, New York 10019*

Brian P. Leaderer

*Department of Epidemiology and Public Health
John B. Pierce Foundation Laboratory
Yale University School of Medicine
New Haven, Connecticut 06519*

INTRODUCTION

Mass concentrations and size-mass distributions of the New York summer aerosol during July and August 1976 were determined by a variety of established and experimental techniques by several of the laboratories collaborating in the New York Summer Aerosol Study (NYSAS). All of the results to be presented and discussed here are based on samples collected, and/or measurements made, at the New York University Medical Center's rooftop sampling station, located at 30th Street near First Avenue, as described by Kneip *et al.*[1]

The objectives of this part of the summer aerosol study were to determine:

1. Temporal concentration patterns of total suspended particulates (TSP), of accumulation mode suspended particulate (AMSP), and of sulfate ion, the most abundant single chemical species within the total suspended particulate.

2. The aerodynamic size distribution of the New York summer aerosol.

3. Sampler performances and reliabilities for the determination of suspended particulate concentrations.

4. Characteristic relations, if any, between the concentration indicated by direct-reading instruments and mass concentrations determined by gravimetric analyses of filter samples.

* Presented May 26, 1977 at the 1977 American Industrial Hygiene Conference, New Orleans, Louisiana.

† Current Affiliations:

Michael T. Kleinman
Rancho Los Amigos Hospital
Downey, California 90242

David M. Bernstein
Brookhaven National Laboratory
Upton, New York 11973

George T. Wolff
General Motors Technical Center
Warren, Michigan 48090

0077–8923/79/0322–0029 $01.75/0 © 1979, NYAS

EXPERIMENTAL METHODS

Sample Collection

Daily samples, running from noon to noon, were collected with the following samplers:

1. A General Metal Works standard high-volume sampler operating at 1.7 m³/min and a constant-flow high-volume sampler, operating at 0.57 m³/min, and consisting of a Roots cycloidal blower with a differential pressure actuated by-pass and a 20 × 25 cm filter holder. The filters used were Gelman Spectrograde glass fibers.

2. A 5-stage Andersen high volume cascade impactor mounted on a General Metal Works Hi-Vol Sampler, and operated at 0.57 m³/min. The sampling substrates on each of' the first four stages were untreated glass fiber filters. Gelman Spectrograde glass fiber filters were used on the final stage.

In addition, two sets of samples were collected on Gelman Spectrograde glass fiber filters with the NYU cyclone-filter type parallel flow particle classifier. These sets were collected for the intervals August 2 through August 9, and August 16 through August 22, respectively.

Gravimetric Analyses

All of the glass fiber filters from the standard Hi-Vol, the Andersen Hi-Vol cascade impactor, and the NYU particle classifier were equilibrated to a temperature of 80° F and relative humidity of 50% and weighed before and after sample collection; a Mettler Gram-atic analytical balance was used to record the weight.

Direct Reading Instruments for Mass Concentration

Two instruments that collect airborne particles from a metered flow of air and measure the accumulated sample mass were provided on loan by their manufacturers and operated during part of the summer study. One was the Thermo-Systems, Inc. Model 3200 Mass Concentration Monitor, which uses electrostatic precipitation to deposit particles onto a quartz-crystal microbalance. The increase in mass on the crystal reduces the frequency of oscillation. The other instrument was a prototype of the GCA/Technology Division's Ambient Particulate Monitor. It collects particles on a filter tape, and measures the mass accumulated on a circle on the filter during the pre-set sampling interval by measuring the change in β-ray attenuation through the circle.

Direct Reading Instruments for Measurements of Other Aerosol Parameters

Knutson, Sinclair and Leaderer [2] described the systems used to measure the diameter distributions of the aerosol and the techniques used to convert them to surface and volume distributions. Volume distributions could be converted to mass distributions if the appropriate corrections for particle density were known. Although they are not, in practice, ever known precisely, the volume distributions can be considered as approximations of the mass distributions.

As discussed by Leaderer, Romano and Stolwijk,[3] the MRI integrating nephelometer measures an integral light-scattering function known as b_{scat}. Various investigators have reported that b_{scat} is proportional to either TSP or AMSP.[4] In fact, MRI provides a conversion factor with the instrument, so that the output can be reported in terms of mass concentration. The manufacturer's conversion factor was used to determine the mass concentration estimates used in this study.

The nephelometer used in this study was equipped with an inlet line drier, and alternate readings were taken 15 minutes with the unmodified ambient aerosol and the aerosol after passing through the drier. The sulfate mass concentrations to be discussed are described in greater detail by Tanner *et al.*[5]

RESULTS

Daily Average Mass Concentration

Measurements of daily TSP concentrations were made between July 12 and September 2, 1976, and are shown in FIGURE 1. FIGURE 2 shows the corresponding measurements of the accumulation mode suspended particulate (AMSP) concentration, based on the samples collected in the last two stages of the Hi-Vol Andersen sampler. FIGURE 3 shows the comparison between the TSP and the nephelometer readings as converted to mass concentration using the manufacturer's calibration. The plot is based on the nephelometer as operated with the inlet drier. FIGURE 4 shows the comparison of the concentration indicated by the nephelometer (with inlet drier) with the AMSP. FIGURE 5 shows the TSP in comparison to the sulfate ion concentration. Finally, FIGURE 6 shows the mass concentration as indicated by the nephelometer (with inlet drier) in comparison to the concentration of sulfate ion. Correlations among selected concentration indices of interest were calculated, and the results are summarized in TABLE 1.

Short-Term Direct Measurements of Mass Concentration

Three instruments providing relatively short interval direct readout measurements of mass concentration were operated during the summer study. The MRI integrating nephelometer and the TSI quartz-crystal mass monitor were operated without apparent problems or obvious malfunctions for most of the study. On the other hand, the GCA β-attenuation mass monitor was only available for about one week in August, and operated according to specification only during parts of three days. It was found to need a more precise regulation of line voltage than was available at the sampling station. The results obtained with the various instruments on August 23–24 are illustrated in FIGURE 7. On this particular day, the nephelometer appeared to indicate a higher concentration than the Hi-Vol TSP. However, as shown in FIGURE 3, the daily average nephelometer concentration readings were lower than the Hi-Vol TSP more often than they were higher.

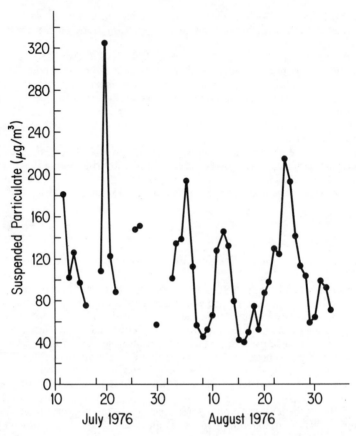

FIGURE 1. 24-hour average Hi-Vol sample TSP concentrations versus start date of sample. The vertical markers indicate Sundays.

The quartz-crystal microbalance instrument gave results that were consistently low in comparison to the Hi-Vol TSP. The β-attenuation mass monitor did not produce a sufficient body of data to permit a realistic evaluation.

Concentration as a Function of Wind Direction

The rapid response of the nephelometer permits its output to be sorted by wind direction. FIGURE 8 shows the average mass concentration as indicated by the nephelometer for each of eight directional segments. The highest concentrations are associated with west, southwest, and south winds, while winds from the northwest, north, northeast and east were associated with the lowest mass concentrations.

Size Distributions

Only two samples were collected with the NYU cyclone-filter, and one of these was incomplete. FIGURE 9 shows the mass concentration as a function of aerodynamic size for the one complete sample, i.e., the one collected between August 16 and August 22, 1976. The figure also shows the size-mass distribution for the Andersen Hi-Vol cascade impactor samples collected during the same time interval, and the particle volume versus diameter distributions for the calculated distributions based on size distribution measurements for the same time interval, which were made with the system described by Knutson, *et al.*[2]

TABLE 2 presents a summary of the size-mass distribution data for the daily Andersen Hi-Vol cascade impactor samples collected during the summer study. The total mass collected on the Anderson impactor averaged 113 μg/m³ with a range of from 43 to 266 μg/m³. Analysis of the stratified mass samples from the 24-hr Andersen runs indicated a mass median diameter of 0.95 μm.

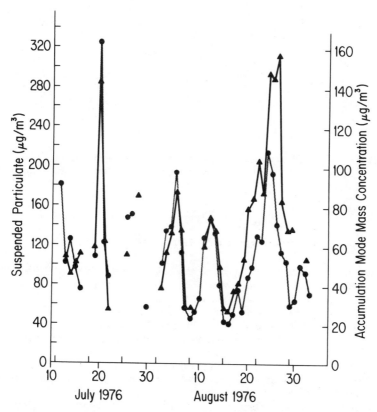

FIGURE 2. Comparison of Hi-Vol TSPs (●) with AMSPs (▲) based on stages 4 and 5 of Hi-Vol Andersen cascade impactor.

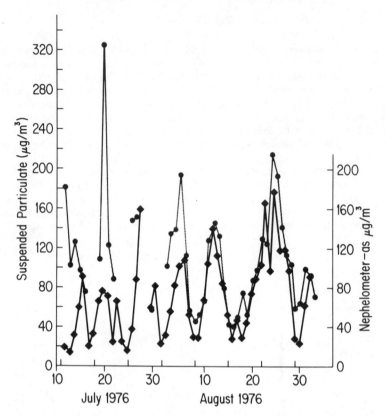

FIGURE 3. Comparison of Hi-Vol TSPs (●) with daily average mass concentrations indicated by an MRI integrating nephelometer operated with an inlet line drier (♦).

DISCUSSION

Cyclic Variations in Daily Average Mass Concentrations

As shown in FIGURES 1, 2, and 5, the daily average levels of TSP, AMSP, and suspended sulfate all exhibited marked cyclic variations, with maxima during the middle of the week, and minima on or close to the weekends. Similar patterns were seen during the same interval by Leaderer, in New Haven, Connecticut,[6] and by Lioy et al. at High Point, New Jersey.[7]

It is interesting that similar patterns of daily average levels of TSP and $SO_4^=$ were observed during the summer of 1975 at High Point, New Jersey, and Whiteface Mt., New York.[8] Hosein et al.[9] made daily measurements of TSP throughout the year of 1973 at Lebanon and Ansonia, Connecticut, and reported that the concentrations were lowest on the weekends.

The immediate cause for the large changes in daily average concentrations is clearly the movement of large-scale weather systems. TABLE 3 was prepared by George Wolff of the Interstate Sanitation Commission, and presents a summary of the weather for the 1976 summer study. It indicates that there were nine cold fronts or changes of air during the 53-day period. This is about one change every six days. Of the nine, six occurred on Saturday or Sunday, and seven occurred on Saturday, Sunday, or Monday.

The last column on TABLE 3, "Expected Pollution," is based on several criteria:

1. Low Pollution—this would be expected to occur during passage of cold fronts, low pressure systems, and during the initial part of a typical high pressure system, especially when NW or NE winds prevail;

2. Moderate Pollution—generally occurs on the second or third day into a new air mass when a westerly flow is becoming established;

3. High Pollution—occurs on backside (SW or W winds) of a high pressure system. This is the part of air mass that has had the longest residence time over U.S. continental sources;

4. Very High Pollution—same as 3 but generally associated with slower moving systems.

While TSP, AMSP, and SO$_4^=$ all show similar trends, it is clear that the greatest excursions during the summer of 1976 occurred for SO$_4^=$. As shown

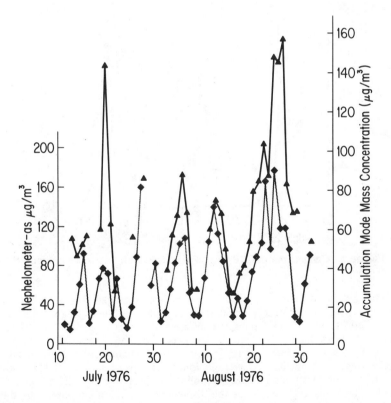

FIGURE 4. Comparison of mass concentration indicated by MRI integrating nephelometer (with inlet drier) (◆) and AMSP (▲).

in FIGURES 5 and 6, the 24-hour maxima were generally in the range of 30 to 40 $\mu g/m^3$, while the minima were generally in the range of 2 to 5 $\mu g/m^3$.

The "Expected Pollution" column on TABLE 3 has a built-in assumption, based largely on relationships between air trajectories and pollution levels. In other words, pollution is expected to be highest when the air mass has passed over heavily populated, polluted regions. The results of the summer study are consistent with this expectation.

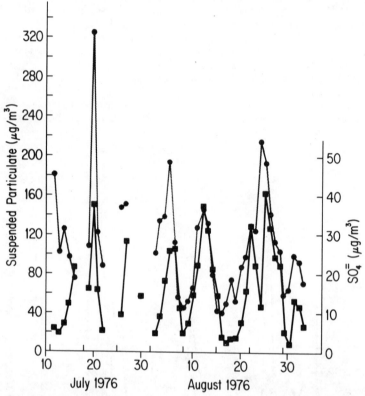

FIGURE 5. Comparison of TSP (●) with sulfate ion mass concentration (■).

On the 17 days labeled "low," the sulfate averaged 8.7 $\mu g/m^3$ (range: 2.1 to 24.1 $\mu g/m^3$); on the 7 "moderate" days, the sulfate averaged 14.9 $\mu g/m^3$ (3.6 to 40.2); on 13 "high" days, the sulfate averaged 18.9 $\mu g/m^3$ (5.2 to 31.4); and on 5 "very high" days the sulfate averaged 32.5 $\mu g/m^3$ (25.4 to 37.3). Using the dispersion index of Kleinman,[10] which incorporates wind speed and mixing depth, the normalized average concentrations of sulfate for low, moderate, high, and very high expected pollution days are 6.0, 27.5, 33.8, and 103.8 $\mu g/m^3$, respectively. Since the dispersion correction eliminates the differences due to dilution volumes and accentuates the differences, it appears that the concentration of sulfate is not heavily influenced by sources within the

TABLE 1

CORRELATIONS AMONG MASS CONCENTRATION INDICES

Variable (x) μg/m³	Variable (y) μg/m³	Number of Pairs	Relationship	Correlation Coefficient (r)*	r² †
TSP (Hi-Vol)	TSP (Andersen)	38	y=1.15x−15.4	0.84	0.71
TSP (Hi-Vol)	Nephelometer (w. drier)	43	y=0.36x+34.37	0.50	0.25
Total SO₄⁼ (Hi-Vol)	Nephelometer (w. drier)	44	y=3.12x+23.6	0.85	0.72
Total SO₄⁼ (Hi-Vol)	Nephelometer (w/o drier)	44	y=3.75x+25.3	0.85	0.72
Accum Mode SO₄⁼ (Andersen)	Nephelometer (w/o drier)	38	y=3.42x+30.4	0.72	0.52

* All of the correlations shown below are significant at the $\alpha=0.001$ level, however, the variation in TSP (Hi-Vol) explains less than 25% of the variation of the nephelometer (w. drier) readings; hence, the relationship between these variables is considered to be tenuous.

† % of y variable variance explained by the x variable.

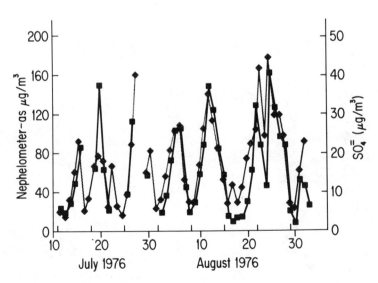

FIGURE 6. Comparison of mass concentration indicated by MRI integrating nephelometer (◆) with sulfate ion mass concentration (■).

TABLE 2

SUMMARY OF DAILY HI-VOL ANDERSEN CASCADE IMPACTOR AERODYNAMIC SIZE DISTRIBUTIONS

	Concentration (in $\mu g/m^3$)					
Stage	5 (Filter)	4	3	2	1	Sum of
Size-Range	<1.1	1.1–2.0	2.0–3.3	3.3–7.0	>7.0 μm	All Stages
Start Date						
July 13	49.2	5.6	11.8	14.0	26.0	106.6
14	41.5	4.0	7.4	10.5	25.3	88.7
15	45.5	5.2	6.2	6.5	19.5	82.9
16	48.0	7.7	3.5	6.3	7.7	73.2
19	54.1	5.6	6.7	11.2	21.6	99.2
20	129.0	12.3	21.5	34.4	68.9	266.1
21	57.1	5.3	7.2	6.6	25.6	101.8
22	26.8	1.7	1.2	9.1	13.0	51.8
26	50.8	4.7	6.5	14.2	28.2	104.4
27	77.4	8.2	8.8	15.9	26.9	137.2
August 2	37.0	1.9	0.9	9.9	21.7	71.4
3	49.8	7.4	6.8	13.6	32.8	110.4
4	62.5	4.4	6.4	18.5	35.2	127.0
5	76.8	10.4	11.7	26.7	36.7	162.3
6	62.1	6.5	8.4	17.0	25.6	119.6
7	28.0	0	0	8.3	10.7	47.8
8	20.9	7.9	4.7	8.9	8.0	50.4
11	50.2	9.3	9.3	16.3	26.1	111.2
12	59.4	15.5	11.6	16.4	37.7	140.6
13	52.3	15.2	11.2	15.4	26.7	120.8
14	42.4	6.8	5.0	6.8	8.1	69.1
15	27.0	1.4	3.9	5.2	5.4	42.9
16	25.9	0.5	5.4	6.5	8.8	47.1
17	37.7	0	8.9	12.1	19.7	78.4
18	35.7	5.8	6.9	13.4	24.2	86.0
19	47.9	5.5	7.9	11.2	27.4	99.9
20	76.9	2.5	10.8	9.1	28.4	127.7
21	74.6	9.9	5.4	17.0	18.2	125.1
22	93.0	10.8	18.3	23.7	40.1	185.9
23	74.2	12.9	13.1	20.9	34.6	155.7
24	144.5	3.0	8.8	16.1	27.2	199.6
25	119.5	25.3	16.4	24.0	44.7	229.9
26	141.0	15.7	13.3	17.9	32.7	220.6
27	69.6	13.4	7.4	9.4	15.0	114.8
28	57.1	11.0	4.6	9.6	13.8	96.1
29	68.3	0.9	11.4	14.9	30.0	125.5
Sept. 1	45.8	8.1	8.6	12.4	22.6	97.5
2	17.4	0	7.0	7.7	12.2	44.3

New York metropolitan region, confirming the analysis of Wolff *et al.*[11] who estimated that up to 85% of the sulfate in the New York City ambient air arrived in the incoming air during the 1976 summer study.

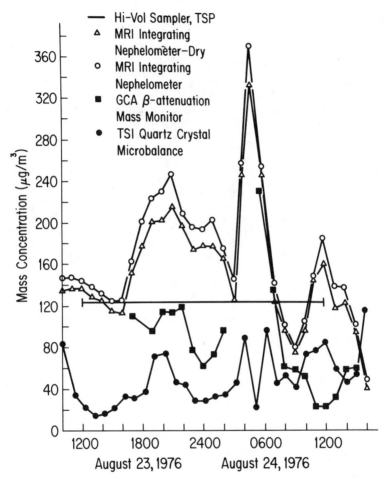

FIGURE 7. Comparison of mass concentrations indicated by gravimetric analysis of Hi-Vol filter, MRI integrating nephelometer (with and without inlet drier), GCA β-attenuation mass monitor, and TSI quartz-crystal microbalance, for noon of August 23 to noon of August 24, 1976.

Utility of the Integrating Nephelometer for Monitoring Suspended Particulate Concentrations

Comparisons of the daily average readings of the nephelometer and TSP, AMSP, and sulfate are shown in FIGURES 3, 4, and 6; and their intercorrelations are presented in TABLE 1. It is clear that the best correlations are between

TABLE 3

NEW YORK SUMMER AEROSOL STUDY WEATHER SUMMARY

Day of Week	Date (1976)	Cold Front	Weather	Expected Pollution *
Sat.	July 10		Back side of polluted air mass	very high
Sun.	July 11		High beginning to move off coast	high
Mon.	July 12	yes	Cold front passes through	low
Tues.	July 13		Brisk NW flow	low
Wed.	July 14		Brisk NW flow	low
Thurs.	July 15		Weak westerly flow established bringing in stagnant air from Midwest	high
Fri.	July 16		New cold front approaching	high
Sat.	July 17	yes	Cold front	low
Sun.	July 18		Westerly flow established	moderate
Mon.	July 19		Backside of high	high
Tues.	July 20		Backside of high	very high
Wed.	July 21		Backside of high	high
Thurs.	July 22		Cold front approaching	high
Fri.	July 23	yes	New air mass (quite small)	low
Sat.	July 24		Front retreating back north and air mass of July 19–21 begins to move back in	moderate
Sun.	July 25	yes	New air mass as new cold front moves through	low
Mon.	July 26		Westerly flow starts	moderate
Tues.	July 27		Backside of high	high
Wed.	July 28		Backside of high	very high
Thurs.	July 29		Backdoor cold front remains just to N of NYC	high
Fri.	July 30		Same as July 29	high
Sat.	July 31		Samle as July 30 but a little closer	moderate
Sun.	Aug. 1	yes	Cold front with developed low	low
Mon.	Aug. 2		New air mass	low
Tues.	Aug. 3		Westerly flow begins	moderate
Wed.	Aug. 4		Backside of high	high
Thurs.	Aug. 5		Backside of high	very high
Fri.	Aug. 6		Cold front approaching	high
Sat.	Aug. 7	yes	Cold front stagnates over NYC	low
Sun.	Aug. 8		Storms develop in front	low
Mon.	Aug. 9		Hurricane Belle	low
Tues.	Aug. 10		Hurricane Belle	low
Wed.	Aug. 11		New air mass which has been sitting over Midwest since Aug. 7 arrives	moderate
Thurs.	Aug. 12		Backside of high	very high
Fri.	Aug. 13		Backside of high	very high
Sat.	Aug. 14		Cold front approaching	high
Sun.	Aug. 15	yes	Cold front	low
Mon.	Aug. 16		New air mass-brisk NW flow	low
Tues.	Aug. 17		New air mass-brisk NW flow	low
Wed.	Aug. 18		New air mass-brisk NW flow	low
Thurs.	Aug. 19		Westerly flow starts	moderate
Fri.	Aug. 20		Westerly flow	moderate
Sat.	Aug. 21		Backside of high	high
Sun.	Aug. 22		Backside of high	very high

TABLE 3—*Continued*

Day of Week	Date (1976)	Cold Front	Weather	Pollution * Expected
Mon.	Aug. 23		Backside of high	high
Tues.	Aug. 24	yes	New air mass as cold front comes through	low
Wed.	Aug. 25		New air mass dissipates and becomes assimilated into air mass of Aug. 23	moderate
Thurs.	Aug. 26		Backside of same high again	high
Fri.	Aug. 27		Disorganized pattern—air masses falling apart, clouds and precipitation develop	low
Sat.	Aug. 28		Same as Aug. 27	low
Sun.	Aug. 29	yes	Cold front	low
Mon.	Aug. 30		New air mass	low
Tues.	Aug. 31		Westerly flow starting	moderate

* Based on position of synoptic scale weather systems.

the nephelometer readings and sulfate, and that the correlation coefficient is essentially equivalent with and without an inlet heater on the nephelometer. The results of this study do not support the use of the nephelometer for monitoring either TSP or AMSP.

The intercomparisons between sulfate ion concentrations and nephelometer readings in TABLE 1 all indicate the relations have strongly positive slopes and positive intercepts. The positive intercepts show that the nephelometer readings are attributable to both sulfate and nonsulfate aerosols. Furthermore, they indicate that the sulfate accounts for most of the light scatter and visibility reduction. The effect of the nonsulfate aerosol on visibility is not very great, but is relatively constant and may be dominant when sulfate levels are low. Grosjean *et al.*[12] have reported that sulfate aerosols are the major species

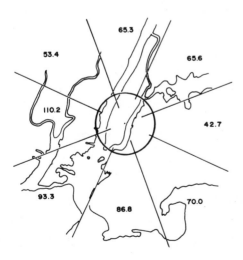

FIGURE 8. Grand average mass concentrations (μg/m^3) indicated by the MRI integrating nephelometer by wind segment for the 1976 summer study at the NYU Medical Center rooftop station.

responsible for visibility reductions in the eastern part of the Los Angeles air basin. Further work is needed at other times of the year and at other locations in order to more firmly establish the role of sulfates in visibility reduction.

Intercomparison of the TSP, AMSP, and Sulfate Concentrations

FIGURE 2 shows that the AMSP was approximately equal to ½ of the TSP during the summer study, except for the interval between approximately August 18 and August 30. As discussed by Wolff *et al.,*[11] there was a large influx of oxidant-rich pollution from the Midwest during this interval, and this may account for the great increase in the fraction of fine particulates.

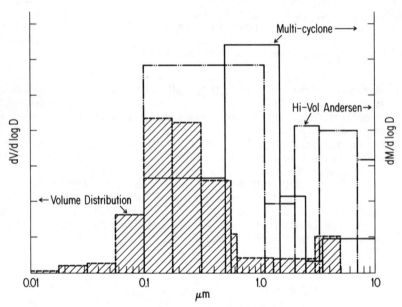

FIGURE 9. Average mass and volume distributions by size intervals for the interval August 16–22, 1976 at the NYU Medical Center rooftop station.

FIGURE 5 shows that the concentration of sulfate ion accounted for up to 25% of the TSP of the New York summer aerosol. Since the sulfate must be associated with cations to form stable chemical compounds, the concentrations of sulfate plus associated cations represents somewhat higher fractions of the TSP. The proportion of sulfate was generally lower for lower values of TSP.

Size Distributions

The three size distributions illustrated in FIGURE 9 are quite different from one another. They all indicate that the distribution is bimodal. The volume distribution exhibits a broad minimum between 0.6 and 3 μm; the Hi-Vol Andersen cascade impactor data indicate a minimum between 1.1 and 2.0 μm, whereas the NYU multicyclone sampler data indicate that the minimum lies between 2.5 and 3.5 μm.

For particles with densities greater than one, the aerodynamic diameters are greater than the linear measures, accounting for some, but not all, of the displacement of the volume histogram.

The Andersen data show much more of the aerosol in the coarse particle mode than either the volume distribution or the multicyclone data. Unfortunately, there are few data available on the size distribution and no standard reference data. Thus, there is no basis at this time for drawing conclusions on the actual size-mass distributions of the New York summer aerosol.

CONCLUSIONS

The daily average TSP, AMSP and sulfate concentrations of the New York summer aerosol follow a strong cyclic pattern. Each cycle lasts approximately one week and the minima generally occur on or just after the weekend, and coincide with the arrival of cold fronts with relatively clean air masses.

The AMSP in the New York summer aerosol generally accounts for about half of the TSP. The sulfate ion, alone, accounts for up to 25% of the TSP. The percent decreases with decreasing TSP. The mass contribution of particles containing sulfate is even greater if one considers that these must also be cations associated with the sulfate ion.

The integrating nephelometer readings correlated well with sulfate concentrations. Its correlations with TSP and AMSP were much weaker. The nephelometer may therefore be more useful as an ambient sulfate monitor than as a TSP or AMSP monitor.

TSP, AMSP, and sulfate concentrations were highest on west, southwest, and south winds. Incoming air, which passed over populated areas had much more TSP, AMSP and sulfate than air passing over less densely populated regions. These differences were increased when corrected for dispersion. Thus, a major portion of the mass and sulfate measured in New York City during the summer study originated in distant upwind regions.

SUMMARY

Size-mass distributions and overall mass concentrations of the New York City summer ambient aerosol were determined by a variety of sampling and analytic techniques. For gravimetric analyses, 24-hour samples were collected daily with high-volume cascade impactors and constant-flow single stage high-volume filters. Seven-day samples were collected with a high-volume, parallel flow particle size classifier, which contains four cyclone-filter two-stage samplers and a total mass filter. One-hour total mass concentration determinations were made with a β-attenuation mass monitor, and one-minute determinations were made with a quartz-crystal microbalance mass monitor. The results of all of the various size-mass distribution and mass concentration measurements are summarized and compared with each other and with estimates based on the measurement of other parameters, i.e., particle volume distributions obtained from an electric aerosol analyzer and optical particle counter, and b_{scat} values obtained with a nephelometer.

ACKNOWLEDGMENTS

The work of New York University is supported by Grant No. RP439–1 of the Electric Power Research Institute and the American Petroleum Institute and is part of a Center program supported by Grant No. ES 00260, from the National Institute of Environmental Health Sciences, and Grant No. CA 13343, from the National Cancer Institute.

Partial support for Dr. Leaderer's research was provided by the Department of Epidemiology and Public Health, Yale University School of Medicine through Grant No. 5-TO1-ES-00123–08, the John B. Pierce Foundation Laboratory Grant No. 5-SO7-RR-05692–07, and the New York State Department of Environmental Conservation.

REFERENCES

1. KNEIP, T. J., B. P. LEADERER, D. M. BERNSTEIN & G. T. WOLFF. 1979. The New York Summer Aerosol Study (NYAS) 1976. Ann. N.Y. Acad. Sci. **322**. (This volume.)
2. KNUTSON, E. O., D. SINCLAIR & B. P. LEADERER. 1979. New York Summer Aerosol: Number Concentration and Size Distribution of Atmospheric Particle. Ann. N.Y. Acad. Sci. **322**. (This volume.)
3. LEADERER, B. P., D. ROMANO & J. A. J. STOLWIJK. 1979. New York Summer Aerosol: Optical Properties. Ann. N.Y. Acad. Sci. **322**. (This volume.)
4. BUTCHER, S. S. & R. J. CHARLSON. 1972. An Introduction to Air Chemistry. Academic Press. New York.
5. TANNER, R., R. GARBER, W. MARLOW, B. P. LEADERER & M. A. LEYKO. 1979. Chemical Composition and Size Distribtuion of Sulfate as a Function of Particle Size in New York City Aerosol. Ann. N.Y. Acad. Sci. **322**. (This volume.)
6. LEADERER, B. P. Personal communication.
7. LIOY, P. J., G. T. WOLFF, K. RAHN, D. M. BERNSTEIN & M. T. KLEINMAN. Characterization of Aerosols Upwind of New York City II. Aerosol Composition. Ann. N.Y. Acad. Sci. **322**. (This volume.)
8. LIOY, P. J., G. T. WOLFF, J. S. CZACHOR, P. E. COFFEY, W. N. STASIUK & D. ROMANO. 1977. Evidence of High Atmospheric Concentrations of Sulfates Detected at Rural Sites in the Northeast. J. Environ. Sci. Health **412**(1, 2): 1–14.
9. HOSEIN, H. R., C. A. MITCHELL & A. BOUHUYS. 1977. Daily Variation in Air Quality. Arch. Environ. Health **32**: 14–21.
10. KLEINMAN, M. T., T. J. KNEIP & M. EISENBUD. 1976. Seasonal Patterns of Airborne Particulate Concentrations in New York City. Atmos. Environ. **10**: 9–11.
11. WOLFF, G. T., P. J. LIOY, B. P. LEADERER, D. M. BERNSTEIN & M. T. KLEINMAN. 1979. Characterization of Aerosols Upwind of New York City I. Transport Processes. Ann. N.Y. Acad. Sci. **322**. (This volume.)
12. GROSJEAN, D. G. J. DOYLE, T. M. MISCHKE, M. P. POE, D. R. FITZ, J. P. SMITH & J. N. PITTS. The Concentration, Size Distribution and Modes of Formation of Particulate Nitrate, Sulfate and Ammonium Compounds in the Eastern Part of the Los Angeles Basin. Presented at the 69th Air Pollution Control Association Meeting, Portland, Oregon, June 1976.

LIGHT-SCATTERING MEASUREMENTS OF THE NEW YORK SUMMER AEROSOL *

Brian P. Leaderer,† David Romano,‡ and Jan A. J. Stolwijk†

† *Department of Epidemiology and Public Health*
John B. Pierce Foundation Laboratory
Yale University School of Medicine
New Haven, Connecticut 06519

‡ *Division of Air Resources*
New York State Department of Environmental Conservation
Albany, New York 12233

INTRODUCTION

Light scattering measurements in a noncloudy atmosphere have been shown to be valuable in assessing visibility, concentrations of the submicron aerosol affecting light scattering (accumulation mode—0.1 μm to 1.0 μm), and, to some extent, the chemical composition of the submicron aerosol itself.[1]

As part of the New York Summer Aerosol Study (NYSAS)[2] light scattering measurements were taken continuously during July and August of 1976 at the New York City site located atop the New York University Medical residence hall at 31st Street on the FDR Drive.

Light extinction measurements were taken at 0.4416 μm utilizing a laser transmissometer system employing a He-Cd laser. The extinction coefficient for scattered light, b_{scat}, as an index for visibility, was measured and the hourly total volume of particles between 0.1 μm and 1.3 μm (the range which most efficiently scatters visible light) were calculated from number size distribution measurements made with an electric aerosol analyzer and optical particle counter.

Light scattering data was analyzed for day-of-week variations, diurnal patterns, effect of wind direction, effect of relative humidity, association with photochemical activity (as measured by ozone) and for relationship with the chemical composition of the submicron aerosol. An intercomparison of light scattering instruments used during the NYSAS will be presented elsewhere.[3]

METHODS

A graphical presentation of the sampling site and the instruments used for measuring the optical properties of the New York summer aerosol are shown in FIGURE 1. The site was located in midtown Manhattan at 47 meters above ground atop the New York University William B. and Cele H. Rubin Hall of Residence at 31st Street on the FDR Drive by the East River. The instruments used and measurements taken were:

* Presented May 26, 1977 at the American Industrial Hygiene Association Conference, New Orleans, Louisiana.

0077–8923/79/0322–0045 $01.75/0 © 1979, NYAS

1. A Meteorology Research Inc. Model 1550 integrating nephelometer equipped with a heated inlet sampling tube and an effective or center wavelength of 0.550 μm was used to measure the point extinction coefficient for scattered light, b_{scat}. The integrating nephelometer was operated continuously from July 12 to September 3, 1976 and hourly average values of $b_{scat} \times 10^{-4}$ m^{-1} were recorded.

2. A long-path low power laser transmissometer system employing a He-Cd laser was used to continuously monitor total extinction ($b_{ext} \times 10^{-4}$ m^{-1}) at a wavelength of 0.4416 μm over a total path of 800 meters. The laser, electronics and receiver optics were located on top of the N.Y.U. residence hall and the retro-reflector was located 400 meters away at the same elevation

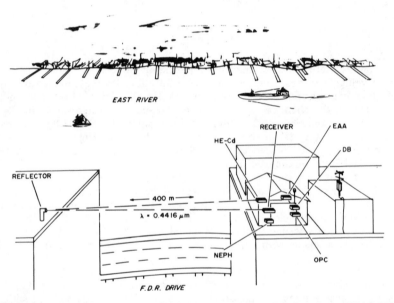

FIGURE 1. Schematic of New York City sampling site for the NYSAS-1976: He-Cd—helium cadium laser transmissometer, EAA-electric aerosol analyzer, DB-Sinclair diffusion battery, OPC-optical particle counter, Neph-integrating nephelometer.

on a building located due North (FIGURE 1). The total extinction coefficient for the laser wavelength was calculated from the percent of transmitted light over the total distance and as such represented a long path integrated measure of light extinction (principally the particulate scattering component). The laser was calibrated by use of neutral density filters. The power output of the He-Cd laser was 15 mw. The He-Cd laser provided approximately 25 days of continuous data, and an average hourly total extinction coefficient for the 0.4416 μm wavelength was recorded.

3. A Sinclair diffusion battery, Thermo Systems Inc. Model 3030 Electric Aerosol Analyser and Royco Instruments Inc. Model 225 optical particle counter with the 518 module were used to obtain hourly aerosol number size

distributions at New York City during the NYSAS and are reported by Knutson *et al.*[4] From a total of 410 complete size distributions recorded, over a total of 13 size ranges from 0.003 μm to 5.0 μm, the total volume for particles in the size range that most directly affects visibility and is primarily responsible for the scattering of visible light (0.1 μm-1.3 μm) was calculated.

4. Total sulfates and nitrates were determined from 24-hour Hi-vol, 24-hour large Andersen 2000 impactor and 12-hour size classified, Sinclair diffusion battery processed samples.[5, 6] Daily sulfate and nitrate data from the Hi-vol and Andersen impactor were collected from July 12 to September 9, 1976 and the diffusion-battery-processed sulfate and nitrate data were collected from August 3 to 26, 1976 (41, 12-hour samples).

5. Daily maximum ozone concentrations (from July 12 to September 9, 1976) were obtained from the New York State monitoring stations located at Roosevelt Island in New York City.

6. A MRI, Inc. Model 1701 weather station provided hourly wind speed, direction and temperature data at the New York City site during August, 1976 and a dew point monitor (cooled mirror principle) located at the New York site provided hourly dew point readings during the same time period. Detailed meteorologic measurements made at the Central park Observatory on an hourly basis were obtained for July and August, 1976 from the National Oceanic Atmospheric Administration.

7. A traffic counter installed on the FDR Drive provided hourly average traffic volumes for the period August 12 to 26, 1976. The diurnal power demand during the summer months for NYC was obtained from Consolidated Edison, New York City's electric power supplier.

RESULTS AND DISCUSSION

Light Scattering—Temporal Variations

Light-scattering measurements made with the integrating nephelometer, b_{scat}, provided a full 8 weeks of continuous hourly data and provided enough data to examine day-of-week variations. The average hourly b_{scat} values were summed over 24-hour periods and averaged by day of week. As shown in FIGURE 2, there was a marked day-of-week variation in light scattering (visibility) recorded during the summer of 1976 in NYC. Initially, it would appear that the average weekly b_{scat} is directly related to the average day-of-week variations in traffic volume recorded on the FDR Drive. However, a closer look at the weather patterns observed during the course of the study showed that a weekly weather cycle existed and was principally responsible for the light scattering pattern observed. The weekly weather cycle is highlighted by the day-of-week averages of wind speed and relative humidity shown in FIGURE 2.

The arrival of a new air mass typically occurred on Mondays and resulted in high wind speeds, low relative humidities and low light scattering or good visibility. As the week progressed wind speeds fell off, the relative humidity increased and light scattering subsequently increased (reduced visibility). Wind speed was lowest, relative humidity highest and light scattering at a maximum on Fridays. The change in the weather systems occurred on Saturdays and

Sundays and resulted in decreasing light scattering. Twenty-four hour total and stratified mass measurements made during July and August as part of the NYSAS showed the same weekly pattern and are discussed by Lippmann *et al.*[7]

Daily maximum oxidant (ozone) concentrations (by chemiluminescent method), obtained from the New York State continuous monitoring station located at Roosevelt Island in New York City for July and August 1976, were averaged by day of week and are shown in FIGURE 2. There was a marked day-of-week variation observed in oxidant concentrations during the summer aerosol study, which followed the light scattering pattern and the observed weekly weather cycle. Oxidant concentrations were observed to be low when

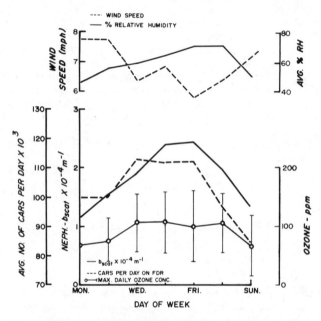

FIGURE 2. Day of week averages for 8 weeks during July and August 1976 for b_{scat}, maximum hourly ozone concentration, % relative humidity and wind speed and day of week averages for 2 weeks during August 1976 for traffic volume in the FDR Drive.

winds were high and humidity low (Sunday, Monday, and Tuesday) and highest when winds are low and humidity high (Wednesday, Thursday, Friday, and Saturday).

Hourly readings recorded continuously during August, 1976 for b_{scat} (by the nephelometer), b_{ext} at $\lambda = 0.4416$ μm (by the HeCd laser transmissometer) and total volume between 0.1 μm and 1.3 μm (by the electric aerosol analyzer and optical particle counter) were averaged by hour of day and are shown in FIGURE 3.

The diurnal pattern for b_{scat}, b_{ext} and total volume between 0.1 μm and 1.3 μm ($V_{.1-1.3}$) exhibited a fairly flat response with little statistically significant hourly variation. $V_{.1-1.3}$ and, to some extent, b_{scat} and b_{ext} showed a small

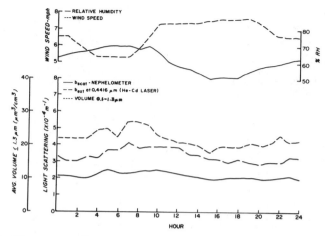

FIGURE 3. The average diurnal pattern for b_{scat}, b_{ext}, total between 0.1μm and 1.3μm ($V_{.1-1.3}$) % relative humidity and wind speed recorded for August 1976 during the NYSAS.

peak in the morning hours corresponding to the morning traffic volume recorded on the FDR Drive (Figure 4) and, to a lesser extent, the daily summer power demand cycle (FIGURE 4). The morning hour peak in $V_{.1-1.3}$, light scattering and total extinction in the New York City summer aerosol is thought to be primarily related to coagulation of aerosol ($<$0.1 μm) emitted from automotive sources during the morning rush hour when atmospheric ventilation is low and humidity is high.

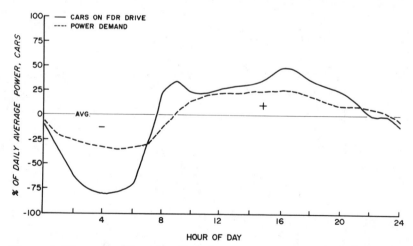

FIGURE 4. The average diurnal traffic and power demand pattern for the summer of 1976 in New York City as percent of the dark average. The traffic data are from counts made on the FDR Drive and the power demand was supplied by Consolidated Eidson and does not include steam generation output.

There were no sharp lows recorded for b_{scat}, b_{ext}, and $V_{.1-1.3}$ during the early morning hours when traffic volume and power demand is lowest. The transport of aerosol, in the 0.1 μm to 1.0 μm size range, into New York City, as discussed by Wolff *et al.*[8] may be primarily responsible for maintaining the levels of b_{scat}, b_{ext}, and $V_{.1-1.3}$ during the early morning hours.

With improved ventilation (wind speed) and reduced relative humidity occurring in the late morning hours there was a slow decline in $V_{.1-1.3}$ and to a lesser extent in b_{scat} and b_{ext}. Aerosol transported into the City and primarily local emissions may be responsible for the afternoon and early evening $V_{.1-1.3}$, b_{scat}, and b_{ext} readings.

Interestingly, there was no observed early afternoon peak in $V_{.1-1.3}$, light scattering or light extinction, indicating that there was no detectable photochemically generated aerosol resulting from local emissions in New York City during the course of the summer study. This finding is in direct contrast to Los Angeles, where a distinct early afternoon peak in $V_{.1-1.3}$ and light scattering was observed and related to photochemical processes resulting from local emissions.[9]

The integrating nephelometer provided the longest continuous record of light scattering (July 12-September 3, 1976) and as such allowed for an analysis of the diurnal light-scattering pattern on days of varying relative humidity. FIGURE 5 shows the diurnal pattern of light scattering, b_{scat}, for days when the relative humidity was in one of four ranges: less than or equal to 50%, 51 to 60%, 61 to 70%, and greater than 70%. The essentially flat average diurnal pattern for light scattering shown in FIGURE 3 was maintained for days of varying relative humidities. However, the magnitude of light scattering increased with increasing relative humidities. The general increase in magnitude of the diurnal light scattering pattern with increasing relative humidities tends to indicate that the general weather cycle (as shown in FIGURE 2) and hence regional air quality and not specific local sources, is largely responsible for the temporal light scattering patterns observed.

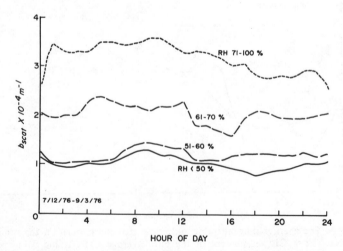

FIGURE 5. The average diurnal pattern for light scattering, b_{scat}, recorded from 7/12/76–9/3/76 for New York City for days when the relative humidity was: less than or equal to 50%; 51 to 60%; 61 to 70%; and greater than 70%.

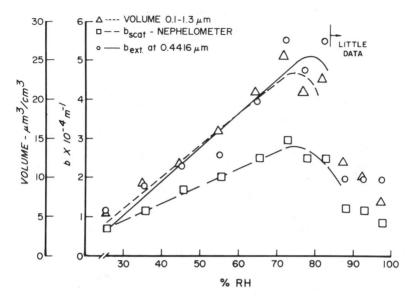

FIGURE 6. Average hourly readings of b_{scat}, b_{ext} and $V_{.1-1.3}$ vs. relative humidity ranges. Ambient temperatures ranged from 10° C to 34° C and dew point from 2° C to 23° C.

Light-Scattering Variations with Meteorologic Variables

Hourly readings of b_{scat}, b_{ext}, and $V_{.1-1.3}$ were compared with relative humidity ranges in which they were measured. A strong correlation was found to exist. With increasing relative humidities there is corresponding increase in b_{scat}, b_{ext}, and $V_{.1-1.3}$ (FIGURE 6). The relationship appears to be linear up to relative humidities of approximately 80% (dew point range of from 8° C to 25° C). For relative humidities above 80% the curves turn downward. However, there were too few recorded readings of b_{scat}, b_{ext}, and $V_{.1-1.3}$ at relative humidity greater than 80% and without rain to evaluate the downward turn.

Significant variations in b_{scat}, b_{ext}, and $V_{.1-1.3}$ were found to exist with wind direction (FIGURE 7). When winds were from the south, southwest and west (the prevailing summer direction) light scattering, extinction and $V_{.1-1.3}$ were observed to be highest, while winds from the northwest, north, northeast, and southeast resulted in significantly lower readings for all three variables. There were too few hours of readings while the winds were from the east to evaluate the effect of that wind direction.

It would appear that elevated relative humidities and wind directions from the west or southwest results in elevated readings of b_{scat}, b_{ext} and $V_{.1-1.3}$ in New York City in the summer, while low relative humidities and winds from the northwest, north-northeast, and southeast result in lower values.

Light-Scattering Variations with Chemical Constituents

In Los Angeles [10, 11] a strong relationship was observed to exist between the evolution of photochemical smog as measured by the maximum ozone concen-

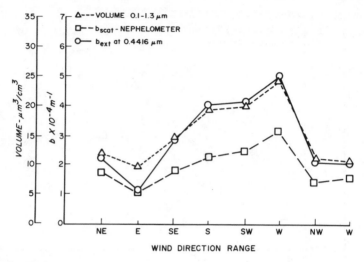

FIGURE 7. Average hourly readings of b_{scat}, b_{ext} and $V_{.1-1.3}$ vs. eight wind direction ranges.

tration and degradation in visibility as measured by light scattering, b_{scat}. This relationship indicated a link between smog chemistry and the heavy haze over the Los Angeles area.

Daily maximum ozone concentrations were obtained from the New York State monitoring station located at Roosevelt Island in New York City. These ozone concentrations were then compared with the light-scattering coefficient for aerosols, b_{scat}, for the same time period to determine if any consistent relationship between aerosol formation, visibility reduction, and evolution of photochemical processes as found in Los Angeles was discernible for the New York City summer aerosol. The results are shown in FIGURE 8.

As can be seen in FIGURE 8 the measured "visibility" decreased systematically (increasing b_{scat}) with increasing maximum daily ozone concentrations. For the 48 observations made during the months of July and August of 1976 in New York City the observed relationship appears to be stronger for New York City than for Los Angeles. However, in New York City there were few occasions where the maximum daily oxidant concentration exceeded 0.2 ppm, whereas the Los Angeles data report several occasions.

The observed relationship between daily maximum oxidant concentrations and b_{scat} in New York City is interesting in view of the flat diurnal patterns recorded for b_{scat}, b_{ext}, and $V_{.1-1.3}$ (FIGURE 3), which indicate no obvious afternoon photochemical aerosol generation. It appears that the observed relationship between daily maximum ozone and b_{scat} in New York City is related to an aged photochemically generated aerosol in the general air mass rather than locally generated photochemical aerosol. That is, the observed relationship between maximum daily ozone concentrations and reduced visibility (increasing b_{scat}) resulted more from the general regional weather pattern and hence regional air quality than specific local New York City sources. The build-up of ozone concentrations and b_{scat} followed a weekly cycle and accumulated under the influence of anticyclonic activity (FIGURE 2) and was consistent with the findings of Wolff *et al.*[12, 13] and Husar *et al.*[14] On days of high ozone, b_{scat} was high all day with no peak occurring at the hours of maximum ozone concentration.

The accumulation mode particles (0.1 μm to 1.0 μm), which are largely responsible for visible light extinction (principally particulate scattering), have been shown to be dominated by a few nonmarine substances, principally sulfates, nitrates, ammonium, and organics.[10, 11] Waggoner *et al.*[15] have indicated that dry particle light scattering, as measured by an integrating nephelometer with dryer, may be closely related to and dominated by ambient sulfate concentrations if a log normal size distribution for the accumulation mode aerosol (0.1 to 1.0 μm) exists and if sulfates account for a substantial portion of the mass of material in the diameter range 0.1 μm to 1.0 μm. Expected SO_4^{-2}/b_{scat} ratios under such conditions would range from 0.1 to 0.2 g-m^{-1}.

The range of sulfate concentrations measured during the NYSAS and reported by Tanner *et al.*[5] was from 2 to 41 μg/m^3. Samples processed by the diffusion battery and samples collected on the large Andersen impactor indicated that about 85% of the sulfate mass was found in respirable particles less than 2 μm in diameter with over 70% in the submicron range and approximately 50% below 0.25 μm. The sulfate species portion of the mass below 2 μm was found, on the average to account for approximately 45% of the total mass below 2 μm. With a large portion of the mass below 2 μm measured during the NYSAS attributable to sulfate compounds and assuming the existance of a log normal size distribution [4] (in the 0.1–1.0 μm range) a good correlation between b_{scat} and sulfate might be expected.

TABLE 1 shows the relationship observed between light scattering, as measured by integrating nephelometer with dryer (b_{scat}) and sulfate measurements made during the NYSAS and reported by Tanner *et al.*[5] The observed relationship was strong for all comparisons made and was independent of the different sampling and analytic methods used to collect and analyze for sulfates.

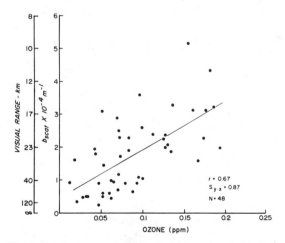

FIGURE 8. Correlation between scattering coefficient for visible light ($b_{scat} \times 10^{-4}$ m^{-1}) and daily maximum ozone concentration (ppm) observed in New York City during the NYSAS 1976.

Correlations coefficients between b_{scat} and sulfate determinations made from 24-hr Hi-vols, 12-hr mass samples ($D \lesssim 5$ μm), and diffusion-battery-processed samples (0.03 μm $\lesssim D \lesssim 5$ μm and 0.3 $\mu \lesssim D \lesssim 5$ μm) were approximately 0.9 or higher indicating that over 80% of the variance of b_{scat} can be associated with sulfate concentrations. The correlation coefficient fell to 0.75 when b_{scat} was compared to sulfates less than 1.1 μm as determined from the backup stage of the large Andersen impactor. The standard error of the estimate, Sy.x, was largest for measurements of total sulfates from the standard Hi-vol and large Andersen impactor and smallest for the diffusion-battery-processed samples.

The average $[SO_4^{-2}]/b_{scat}$ ratio for the New York summer aerosol was found to be from approximately 0.05 to 0.07 gm^{-2} for sulfates in the size range

TABLE 1

MEASURED RATIOS AND CORRELATIONS OF SO_4^{-2}, AND b_{scat}

$[SO_4^{-2}]$ determined from	Number of Samples	$[SO_4^{-2}]/$ b_{scat} (gm^{-2})	$(\pm\sigma)$*	SO_4^{-2} vs. b_{scat} r	SO_4^{-2} vs. b_{scat} Sy·x $(\mu g/m^3)^c$
24 hour Hi-Vols	48	0.08	($\pm$ .03)	0.89	$\pm$5.1
Large Anderson Impactor $D \lesssim 1.1\mu$m	37	0.07	($\pm$ .03)	0.75	$\pm$5.8
Diffusion sampler ‡ un-treated air $D \lesssim 5\mu$m	40	0.053	($\pm$0.018)	0.92	$\pm$3.99
Diffusion sampler 0.03μm $\lesssim D \lesssim 5\mu$m	41	0.052	($\pm$0.018)	0.94	$\pm$3.63
Diffusion sampler 0.3 $\lesssim D \lesssim 5\mu$m	40	0.027	($\pm$0.011)	0.91	$\pm$2.27

All correlations significant at the 0.001 level.
* Standard deviation.
† Standard error of the estimate.
‡ 12-Hour samples.

0.1 to 1.0 μm (TABLE 1). This value is less than the theoretical and measured value of $\sim$0.1 gm^{-2} and suggests that approximately half the optical effects observed during the NYSAS could be attributed to sulfates.

The sensitivity of light scattering to sulfate concentrations in the optical and suboptical size ranges would tend to suggest that the size distribution of the sulfate aerosol in the New York summer aerosol remained fairly constant in shape. Additional multiple regression analysis employing measured nitrate values failed to improve the explained variance in b_{scat}. This failure was due principally to the fact that nitrates were found to constitute a small fraction of the total submicron mass (approximately 2%) and as such would contribute little to light scattering. Unfortunately there were few data available on the organic component of the submicron mass making it impossible to assess the organic component's contribution to light scattering.

The relationship between b_{scat} and sulfates observed in New York City during the NYSAS was found to be much stronger than that observed in Sweden by Waggoner *et al.*[14]

The strong observed effects of sulfates on light scattering in the New York City summer aerosol is particularly interesting in view of the poor relationship observed between light scattering and total mass and submicron mass measured during the NYSAS and reported by Lippmann *et al.*[7] There is a clear need for additional data, collected during different seasons in urban and rural areas, which will better define the relationship between the submicron mass (its physical and chemical properties, principally sulfates) and light scattering.

Conclusions

A strong relationship was observed between light scattering, as measured by an integrating nephelometer with dryer, and water soluble sulfate indicating that sulfates dominated light scattering in the New York summer aerosol.

Light scattering and ozone concentrations were found to follow a daily pattern which was related to the weekly weather cycle that existed during the course of the study. A flat diurnal pattern was observed for light scattering, light extinction and the aerosol volume between 0.1 and 1.3 μm with a small peak occurring at 8:00 A.M. and thought to be related to automotive sources. The flat diurnal pattern was consistent when analyzed for days of varying relative humidity.

Elevated relative humidities and wind directions from the west or southwest resulted in elevated readings of b_{scat}, b_{ext}, and $V_{.1-1.3}$ in New York City in the summer, while low relative humidities and winds from the northwest, north-northeast and southeast result in lower values.

A positive relationship between maximum daily ozone concentration and the corresponding light scattering was found, suggesting some photochemical generation of aerosol resulting from local emissions. However, no afternoon peak in light-scattering measurements or $V_{.1-1.3}$ was found (for all days averaged or on a day to day basis), suggesting rather that visibility reduction (increasingly light scattering) and ozone concentrations in NYC are the result of the aged, general air mass affecting the New York City region rather than the result of photochemical processes resulting from local emissions.

In summary it would appear that elevated light scattering in the New York City summer aerosol is strongly related to stagnant air masses arriving from the west-southwest and containing a photochemically produced aged aerosol (principally sulfate).

Acknowledgments

Partial support for this work was provided by the Department of Epidemiology and Public Health, Yale University School of Medicine through Grant No. 5-T01-ES-00123–08, the John B. Pierce Foundation Laboratory through Grant No. 5-S07-RR-05692–07, and the New York State Department of Environmental Conservation.

REFERENCES

1. WAGGONER, A. P. & R. J. CHARLSON. 1976. Measurements of Aerosol Optical Parameters. *In* Fine Particles. B. Y. H. Liu, Ed. Academic Press. New York.
2. KNEIP, T. J., LEADERER, B. P., D. M. BERNSTEIN & G. T. WOLFF. 1979. New York Summer Aerosol Study (NYSAS)—1976. Ann. N.Y. Acad. Sci. **322.** (This volume.)
3. LEADERER, B. P. & J. A. J. STOLWIJK. Comparison of Optical Monitors Used in the New York Summer Aerosol Study (NYSAS). In preparation.
4. KNUTSON, E. O., D. SINCLAIR & B. P. LEADERER. 1979. New York Summer Aerosol: Number Concentration and Size Distribution of Atmospheric Particles. Ann. N.Y. Acad. Sci. **322.** (This volume.)
5. TANNER, R., R. GARBER, W. MARLOW, B. P. LEADERER & M. A. LEYKO. 1979. Chemical Composition and Size Distribution of Sulfate as a Function of Particle Size in the New York City Summer Aerosol. Ann. N.Y. Acad. Sci. **322.** (This volume.)
6. KLEINMAN, M. T., B. P. LEADERER, C. TOMACZK & R. TANNER. Inorganic Nitrogen Compounds in the New York City Summer Aerosol. Ann. N.Y. Acad. Sci. **322.** (This volume.)
7. LIPPMANN, M., M. T. KLEINMAN, D. M. BERNSTEIN, GEORGE T. WOLFF & B. P. LEADERER. 1979. Size-Mass Distributions of the New York Summer Aerosol. Ann. N.Y. Acad. Sci. **322.** (This volume.)
8. WOLFF, G. T., P. J. LIOY, B. P. LEADERER, D. M. BERNSTEIN & M. T. KLEINMAN. Characterization of Aerosols Upwind of New York City, II. Aerosol Composition. Ann. N.Y. Acad. Sci. **322.** (This volume.)
9. WHITBY, K. T., R. B. HUSAR & B. Y. H. LIU. 1972. The Aerosol Size Distribution of Los Angeles Smog. J. Colloid Interface Sci. **39:** 177.
10. HIDY, G. M. 1974. Characterization of Aerosols in California (ACHEX). Vol. 2 Summary. (Prepared for California Air Resources Board). Rockwell International. Thousand Oaks, Cal.
11. HIDY, G. M. 1975. Summary of the California Aerosol Characterization Experiment. J. Air Pollut. Control Assoc. **25:** 1106.
12. WOLFF, G. T., P. J. LIOY, B. P. LEADERER, D. M. BERNSTEIN & M. T. KLEINMAN. Characterization of Aerosols Upwind of New York City, I. Transport. Ann. N.Y. Acad. Sci. **322.** (This volume.)
13. WOLFF, G. T., P. J. LIOY, G. D. WRIGHT, R. E. MEYERS & R. CEDERWALL. 1977. An Investigation of Long-Range Transport of Ozone Across the Midwestern and Northeastern U.S. Atmos. Environ. **11:** 797–802.
14. HUSAR, R. B., D. E. PATTERSON, C. C. PALEY & N. V. GILLANI. 1977. Ozone in Hazy Air Masses. Paper No. 6–5 . *In* Proceedings of the International Conference on Photochemical Pollution U.S. E.P.A. Pub. No. EPA–600/3–7–001b. Research Triangle Park, N.C.
15. WAGGONER, A. P., A. J. VANDERPOL, R. J. CHARLSON, S. LARSEN, L. GRANAT & C. TRAGARDH. 1976. Sulfate—Light Scattering Ratio as an Index of the Role of Sulfur in Tropospheric Optics. Nature **261:** 121.

CHARACTERIZATION OF AEROSOLS UPWIND OF NEW YORK CITY: I. TRANSPORT *

George T. Wolff † and Paul J. Lioy †

Interstate Sanitation Commission
New York, New York 10019

Brian P. Leaderer

Department of Epidemiology and Public Health
John B. Pierce Foundation Laboratories
Yale University School of Medicine
New Haven, Connecticut 06519

David M. Bernstein † and Michael T. Kleinman †

Institute of Environmental Medicine
New York University Medical Center
New York, New York 10016

INTRODUCTION

In order to characterize the aerosol generated within the New York City-northeastern New Jersey Metropolitan Area, it was necessary to characterize the background, or upwind, aerosol that contributes to some fraction of the observed total suspended particulates (TSP) in New York City. A site in rural western N.J. at High Point State Park was chosen for the background site because of the predominance of westerly winds in this area. In Part II [1] of this characterization (see next paper), the aerosol composition will be discussed. In this paper, transport is examined.

High Point is situated near the common border of New Jersey, New York, and Pennsylvania, and is 55 miles west-northwest of the New York city site, 60 miles northeast of the Allentown-Bethlehem area and approximately 50 miles east of Scranton (see FIGURE 1). The monitoring site (FIGURE 2) was situated on a 15-mile long north-south ridge at the 1640 ft. level. Immediately to the west and east, the elevation drops off rapidly to about 400–500 ft. The area surrounding High Point is sparsely populated, with a population density

* Revised Version of the paper presented at the American Industrial Hygiene Association 1977 Conference, New Orleans, Louisiana, May 22–27, 1977, as part of the New York Summer Aerosol Study (NYSAS).

† Current Affiliations:

George T. Wolff
Environmental Science Department
General Motors Research Laboratories
Warren, Michigan 48090

Paul J. Lioy
Institute of Environmental Medicine
New York University Medical Center
New York, New York 10016

David M. Bernstein
Brookhaven National Laboratory
Upton, New York 11973

Michael T. Kleinman
Rancho Los Amigos Hospital
University of Southern California
Downey, California 90242

0077–8923/79/0322–0057 $01.75/0 © 1979, NYAS

FIGURE 1. Map showing location of High Point with respect to New York City.

of less than 100 people per square mile, and contains no large point sources. Consequently, with west-northwest winds, High Point is an ideal site for observing contaminants from upwind of New York City.

PROCEDURES

Twenty-four simultaneous high volume samples were collected daily at High Point and at the New York University Medical Center in New York City for two alternate weeks. The samples were obtained on a noon-to-noon basis. Filters were weighed for TSP and subsequently analyzed for trace metals and water soluble sulfates. The procedures have been discussed elsewhere.[2]

RESULTS

Total Suspended Particulates and Sulfates

The TSP and sulfate data obtained during the period are shown in FIGURES 3 and 4. The mean TSP concentrations at High Point and New York City during the study were 65.5 and 93.8 $\mu g/m^3$, respectively. Mean $SO_4^=$ concentrations were 8.83 at High Point and 9.90 $\mu g/m^3$ at New York City.

Air Parcel Trajectories

Backward air parcel trajectories were calculated for each day from the surface to 1000 m. Trajectories terminate in New York City at 0100h, which is in the middle of the 24h sampling period. However, on Aug. 6, a 0700h trajectory was used because of insufficient 0100h wind data. Calculation procedures have been described elsewhere.[3]

On days with sulfate concentrations greater than 7.5 $\mu g/m^3$, the trajectories are shown in FIGURE 5. For days with sulfate levels less than 7.5 $\mu g/m^3$, the trajectories are shown in FIGURE 6.

The path of a particular air parcel is denoted by a solid, dotted, or dashed line. The number in parentheses at the beginning of the trajectory is the date in August, 1976 when the trajectory terminated. The numbers along the

FIGURE 2. Aerial view of the region immediately around the monitoring site at High Point.

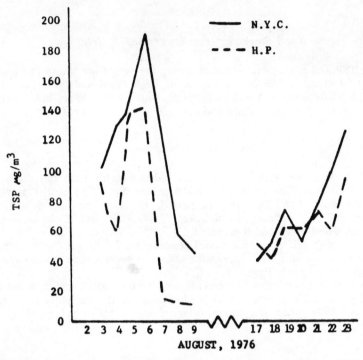

FIGURE 3. The total suspended particulate concentrations observed at High Point, N.J., and New York City during the NYSAS.

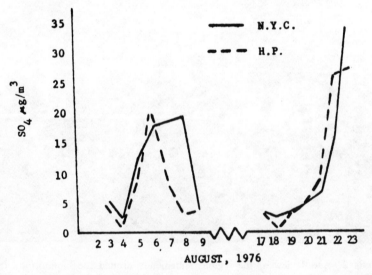

FIGURE 4. The sulfate concentrations observed at High Point, N.J. and New York City during the NYSAS.

FIGURE 5. Trajectories for days with sulfate values greater than 7.5 μg/m³.

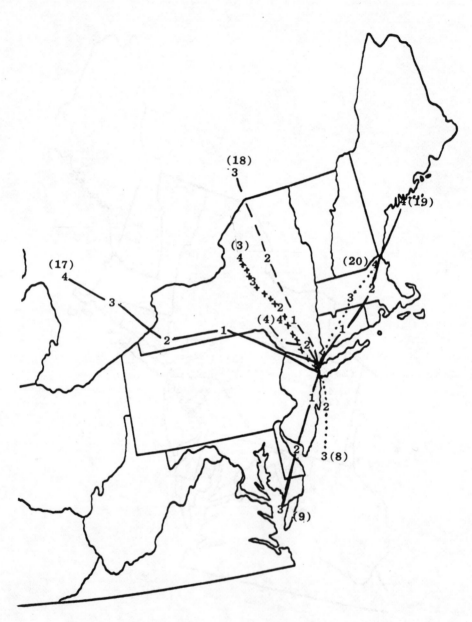

FIGURE 6. Trajectories for days with sulfate values less than 7.5 μg/m³.

trajectory path indicate the previous position in 6 hour intervals. For example, the number 1 indicates the position 6 hours previous to the parcel's arrival in New York. The number 2 is the position 12 hours prior to termination, etc.

It appears that the higher sulfate values are associated with trajectories arriving from the west-northwest. The difference between the Aug. 17 west-northwest trajectory, which produced low sulfates, and the others in FIGURE 5, which resulted in high sulfates, was probably a result of the wind speed. The wind speed on Aug. 17 was over two times the speed on Aug. 6, 7, 22, and 23.

Transport

Particles with a diameter of less than or equal to 2 μm have a residence time in the atmosphere on the order of days.[4] Photochemically generated aerosol is found in this size range. Miller *et al.*[5] sampled the New York City aerosol for 17 days during Aug. 1972 and found that between 26–59% of the total aerosol was particles with diameters less than 2 μm. Results obtained in this present study [6] indicate that 50% of the mass was composed of particles less than 2 μm in diameter. During the period of Aug. 22 and 23, approximately 80% of the mass was contained in particles less than 2 μm. According to Miller *et al.*,[5] these particles consist primarily of sulfates, ammonium, and organic compounds. The fraction of sulfates less than 2 μm is approximately 85%.[5, 6] All of these constituents are associated with photochemical smog.[7-11] The simultaneous occurrence of elevated sulfates and ozone during periods of high photochemical smog in New York and New Jersey has previously been documented.[1, 12, 13] Consequently, sulfates should be a fairly good indicator of the presence of photochemically generated aerosols less than 2 μm in diameter.

From the trajectories in FIGURES 5 and 6, it is evident that High Point was directly upwind from New York City on Aug. 6, 7, 17, 21–23. On Aug. 21, however, the trajectory from High Point headed south toward the Philadelphia area before turning northeastward toward New York City. Since the parcel could have picked up contaminants from the Philadelphia area, it will not be included in the following discussions.

The comparative upwind and downwind data are shown in TABLE 1. Since the trajectories terminate in New York City during the middle of the sampling period, and since all of the trajectories except the one on Aug. 18 indicate about a 12-hour travel time from High Point to New York City, the upwind values on the day the trajectory terminates and the previous day must be considered. As a result, the average concentration observed at High Point during these two days was compared to the New York City value on the second day. On Aug. 17, the concentrations were compared directly because the travel time of the air parcel from High Point to New York City was about 3 hours.

The ratio of TSP values at High Point to New York City ranged from 0.59 to 0.86 (avg. = 0.69) and for SO$_4^=$ the ratio ranged from 0.76 to 1.13 (avg. = 0.86). These data suggest a consistent relationship between the air at High Point and the air in New York City. Since the environment near High Point is very rural, the observed concentrations cannot be accounted for by local sources. As a result, a major portion of both the observed TSP and SO$_4^=$ at High Point is probably due to long-range transport. In addition, considering the long lifetime of the sulfate aerosol, the data suggest that as much as 73%

TABLE 1

COMPARISON OF SIMULTANEOUS UPWIND (HIGH POINT) AND DOWNWIND (NEW YORK CITY) TSP AND SO₄ CONCENTRATIONS ON DAYS WITH WEST TO NORTHWEST WINDS

	Upwind (High Point)					Downwind (NYC)		Ratio	
	TSP ($\mu g/m^3$)		$SO_4^=$ ($\mu g/m^3$)			TSP ($\mu g/m^3$)	$SO_4^=$ ($\mu g/m^3$)	$\dfrac{TSP_{H.P.}}{TSP_{NYC}}$	$\dfrac{SO_{4\,H.P.}}{SO_{4\,NYC}}$
Date	Daily	Mean	Daily	Mean	Date				
8/5	140	141.5	7.9	14.2	8/6	194	18.3	.73	.78
8/6	143		20.5						
8/6	143	78.5	20.5	14.4	8/7	112	19.0	.70	.76
8/7	14		8.3						
8/18		43		1.9	8/18	50	2.5	.86	.76
8/21	75	66	9.0	17.6	8/22	100	15.6	.66	1.13
8/22	57		26.2						
8/22	57	76	26.2	26.7	8/23	129	32.0	.59	.83
8/23	95		27.2						
Mean		81		15.0		117	17.5	.69	.86

(86% $SO_4^= \times$ 85% < 2 μm) of the sulfates observed in New York City could be accounted for by transport through High Point on those days when the air parcels passed through High Point before arriving in New York City. The data also indicate that a significant portion of the TSP observed at High Point reached New York City. At higher wind speeds, the TSP at High Point constitutes a greater fraction of the TSP in New York City (TABLE 2). This would be expected because with higher wind speeds, the time of transport to New York City would be less so that particles would have less time to fall out and the turbulence would allow the larger particles to remain airborne longer.

Since the average ratio of TSP at High Point to New York City is 0.69 and the average fraction of TSP less than 2 μm is 0.50, approximately 34% of the mass could have passed through High Point en route to New York. During the period of August 22 and 23, when 80% of New York City's TSP was below 2 μm in diameter, as much as 74% of the total could have been due to long-range transport.

TABLE 2

THE RATIO OF TSP AT HIGH POINT TO TSP AT NEW YORK CITY RANKED ACCORDING TO THE WIND SPEED FROM THE SURFACE TO 1000 m

Date (ranked according to increasing wind speed)	$\dfrac{TSP_{H.P.}}{TSP_{NYC}}$
8/22–8/23	.59
8/21–8/22	.66
8/6–8/7	.70
8/5–8/6	.73
8/17	.86

Relationship between TSP & SO₄ Episodes and the Arrival of Areas of Low Visibility and High Ozone

Husar *et al.*[14] demonstrated that visibility measurements can serve as tracers of hazy summertime air masses containing high $SO_4^=$ concentrations. Subsequently, Wolff *et al.*[15] and Husar *et al.*[16] demonstrated that these hazy air masses also contained elevated ozone concentrations, which indicated a relationship between low visibility, sulfate, and intense periods of photochemical activity.[10] Recently, Weiss *et al.*[17] found that sulfate particles dominated the submicrometer size, light-scattering component of the aerosol at several rural midwestern and southern sites. This suggested that sulfate occurs over large geographic regions. The uniformity of the observed sulfate concentrations over rural and urban areas suggest that summertime sulfate is not solely due to local emissions. In addition, Lippmann[6] found an excellent correlation in New York City between b_{scat} obtained with an integrating nephelometer and soluble sulfate mass. Consequently, visibility measurements or nephelometer data should serve as an indicator of sulfates, and both low visibility and high ozone should serve as indicators of air masses with significant photochemical aerosol production.

FIGURE 7 contains visibility and ozone maps for the period Aug. 19–23.

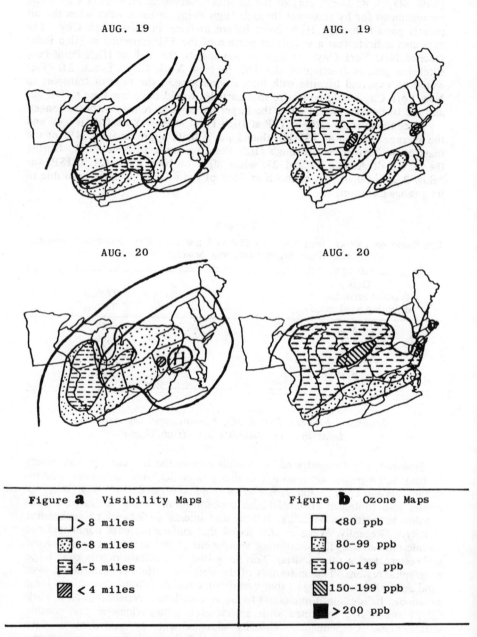

FIGURE 7. a. Visibility isopleths.
b. Ozone isopleths for August 19 through 23, 1976.

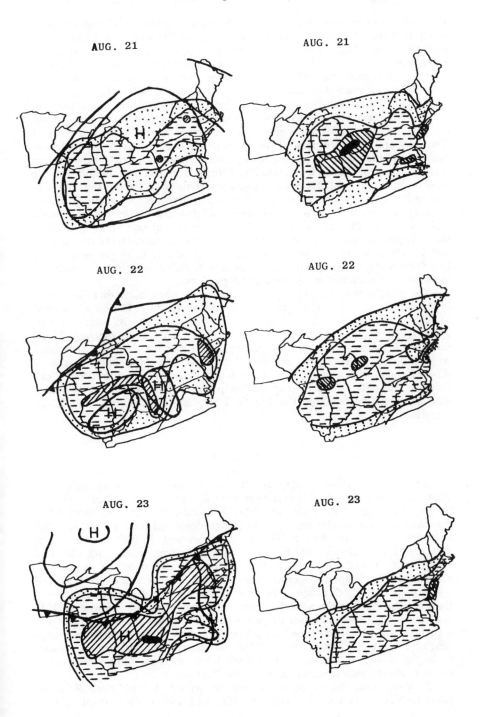

The visibility maps were generated using National Weather Service observations obtained every three hours. The data appearing on the maps are the minimum daily visibilities observed in the absence of precipitation or fog and with a dew point depression of greater than 2° F. This eliminated any visibility reductions resulting from moisture rather than pollution. Ozone maps were constructed from daily maximum data obtained from the 19 states within the area.

FIGURE 7 illustrates the concomitant behavior of high ozone and low visibilities. Although not shown, visibility maps were also plotted for Aug. 16–18. On Aug. 16, an area of low visibility formed over most of Ohio and parts of Kentucky and West Virginia. This air mass remained relatively stationary on Aug. 17 as little movement of the high pressure system occurred. However, on Aug. 17, a second area of low visibility formed over much of Illinois and the western shore of Lake Michigan. The geographic areas where the low visibilities first began to appear are the same areas where elevated ozone levels are typically first observed in a new air mass that has just moved out of Canada and are located over or downwind of high density emission areas.[18] Both of these air parcels containing low visibility and high ozone observed on Aug. 16 and 17 intensified and expanded on Aug. 18. By the 19th, as the high moved into New York State, the two areas of low visibility merged and expanded eastward into Western Pennsylvania and New York.

The Aug. 22 trajectory (FIGURE 5) supports the observation that the area of low visibility over Ohio and Western Pennsylvania on Aug. 19 and 20 reached the East Coast on Aug. 21. This coincided with the onset of increased TSP and $SO_4^=$ values observed at High Point and New York City on the Aug. 21–22 samples. Similarly, the Aug. 23 trajectory shows that the air parcel associated with visibilities of less than 4 miles over the New York Metropolitan Area on Aug. 22 corresponded to the area of 4–5 mile visibility over Western New York State on Aug. 21. The arrival of this air parcel in New York City coincided with additional increases in the TSP and $SO_4^=$ observed in the Aug. 22–23 samples.

DISCUSSION

High concentrations of $SO_4^=$ and TSP have been observed at rural High Point, N.J., and these high levels cannot be accounted for by local emissions or only by short-range transport. The data indicate that both parameters generally followed similar trends at both High Point and New York City. One exception occurred on Aug. 8. On this day, they were in different air masses as a northeast-southwest cold front was located between New York City and High Point. Trajectory analysis indicates that on several of the days which experienced the highest concentrations, the air flow over the previous 24h was from a westerly direction. The onset of the periods of elevated concentrations corresponded to the arrival of hazy air parcels from the Midwestern U.S. containing high concentrations of photochemical oxidants. Since sulfates and a portion of the total particles have lifetimes of days in the atmosphere,[4] long-range transport of these pollutants from midwestern industrial centers appears to be the source of elevated contaminant levels at High Point and a major factor in the elevated concentrations in New York City.

On days when the air parcel trajectories passed through High Point on route to New York City, the ratios of $SO_4^=$ and TSP at High Point to New

York City averaged 0.86 and 0.69, respectively. This means that as an average of 73% and a maximum of 86% of the $SO_4^=$ observed in New York City originated upwind of High Point. These data indicate that a substantial portion of the TSP measured in New York City on these days also originated upwind of High Point. This portion averaged 34% and went as high as 47%.

It should be noted that the accuracy of these numbers is largely dependent upon the accuracy of the trajectory calculations, especially during conditions conductive for nocturnal inversion development which occurred during each of the sampling periods classified as high sulfate days. In a previous paper,[3] it was estimated that the mean displacement error of a given trajectory is about a nautical mile per hour of travel time. During a nocturnal inversion when approximately the lowest 100m is decoupled from the geostrophic flow, the displacement error of a 1000m-thick parcel does not seem to increase significantly for two reasons. First, the thickness of the nocturnal boundary layer (100m) is relatively small compared to 1000m. Second, the diminished winds in the lowest 100m are usually partially offset by increased winds within the nocturnal jet,[19] which develops in a layer several hundred meters thick just above the inversion layer.

Although the mean displacement error of an air parcel 1000m thick should not increase significantly during a nocturnal inversion, the path of emissions from sources trapped below the nocturnal inversion layer will not be described by the 0–1000m trajectory. However, since the primary source of summertime sulfates in the northeastern U.S., appears to be from elevated power plant plumes,[20] most of the sulfur is emitted above the nocturnal inversion layer.

CONCLUSIONS

The above analysis indicates that on certain summer days, most of the sulfate observed in New York City was due to long-range transport from sources in the midwestern U.S. It is estimated that the average fraction of $SO_4^=$ due to transport is 0.73. The above analysis also indicates that a significant fraction of the TSP observed in New York City on high pollutant summer days is a result of long-range transport.

The occurrence of high $SO_4^=$ and TSP levels in the summer in New York City is associated with the incursion of hazy air masses containing high levels of photochemical oxidants which have been transported from the midwestern U.S.

ACKNOWLEDGMENTS

The authors are grateful to the members of the Moodus Ozone Data Analysis Task Force for providing the daily ozone data, to Mr. Richard Cederwall and Dr. Ronald Meyers of Brookhaven National Laboratory for providing the air parcel trajectories, and to Dr. Ralph Pasceri of the New Jersey Department of Environmental Protection for providing some of the instrumentation at High Point. In addition, the authors thank Mr. William Bailey of High Point State Park and Messrs. William Edwards and Konrad Wisniewski for their assistance in this study.

REFERENCES

1. LIOY, P. J., G. T. WOLFF, K. A. RAHN, D. M. BERNSTEIN & M. KLEINMAN. 1979. Characterization of Aerosols Upwind of New York City, II. Aerosol Composition. Ann. N.Y. Acad. Sci. 322. (This volume.)
2. KNEIP, T. J., B. LEADERER, D. M. BERNSTEIN & G. T. WOLFF. 1979. The New York Summer Aerosol Study (NYSAS), 1976. Ann. N.Y. Acad. Sci. 322. This volume.
3. WOLFF, G. T., P. J. LIOY, R. E. MEYERS, R. T. CEDERWALL, G. D. WIGHT, R. E. PASCERI & R. S. TAYLOR. 1977. Anatomy of Two Ozone Transport Episodes in the Washington, D.C. to Boston, Mass. Corridor. Environ. Sci. Technol. 11: 506.
4. WILLEKE, K. & K. T. WHITBY. Physical Characteristics of Denver Area Aerosols. Paper No. 74–262, In Proceedings of the 67th Meeting of the Air Pollution Control Assoc. Denver, Colo., June 9–13, 1974.
5. MILLER, D. F., W. E. SCHWARTZ, P. W. JONES, D. W. JOSEPH, C. W. SPICER, C. J. RIGGLE & A. LEVY. 1973. Haze Formation: Its Nature and Origin. U.S. EPA Publ. No. EPA-650/3-74-002, Research Triangle Park, N.C., June.
6. LIPPMANN, M., M. KLEINMAN, D. M. BERNSTEIN, G. T. WOLFF & B. LEADERER. 1979. Size-mass distributions of the N.Y. Summer Aerosol. Ann. N.Y. Acad. Sci. 322. (This volume.)
7. HIDY, G. M. 1975. Summary of the California Aerosol Characterization Experiment. J. Air Pollut. Control Assoc. 25: 1106.
8. MUELLER, P. K., R. W. MOSLEY & L. B. PIERCE. 1972. Chemical Composition of Pasadena Aerosol by Particle Size and Time of Day. In Aerosols and Atmospheric Chemistry. G. M. Hidy, Ed. Academic Press, New York. pp. 295–299.
9. GROSJEAN, D., G. J. DOYLE, T. M. MISCHKE, M. P. POE, D. R. FRITZ, J. P. SMITH & J. N. PITTS. The Concentration, Size Distribution and Modes of Formation of Particulate Nitrate, Sulfate and Ammonium Compounds in the Eastern Part of the Los Angeles Basin. Paper No. 76–20.3, In Proceedings of the 69th Meeting of the Air Pollution Control Assoc., Portland, Oregon, June 1976.
10. BROSSET, C. 1976. Air-borne Particles: Black and White Episodes. Ambio 5: 157.
11. BURTON, C. S., T. N. JERSKEY & S. D. REYNOLDS. 1977. A Preliminary Investigation of Expected Visibility Improvements in the L.A. Basin from Oxidant Precursor Gases and Particulate Emission Controls. Paper No. 9–4, In Proceedings of the International Conference on Photochemical Pollution. U.S. EPA Pub. No. EPA-600/3-77-001, Research Triangle Park, N.C.
12. STASIUK, W. N., P. E. COFFEY & R. F. McDERMOTT. 1975. Relationships between Suspended Particulates and Ozone at a Non-urban Site. Paper No. 75–62.7, In Proceedings of the 67th Meeting of the Air Pollution Control Assoc., Boston, Mass., June 1975.
13. LIOY, P. J., G. T. WOLFF, J. S. CZACHOR, P. E. COFFEY, W. N. STASIUK & D. ROMANO. 1977. Evidence of High Atmospheric Concentrations of Sulfates Detected at Rural Sites in the Northeast. J. Environ. Sci. Health Part A, 12: 1.
14. HUSAR, R. B., J. D. HUSAR, N. V. GILLANI, S. B. FULLER, W. H. WHITE, J. A. ANDERSON, W. M. VAUGHAN & W. E. WILSON, JR. Pollutant Flow Rate Measurements in Large Plumes: Sulfur Budget in Power Plant and Area Source Plumes in the St. Louis Region. Presented at the 172nd American Chemical Society Meeting, New York, N.Y., April 1976.
15. WOLFF, G. T., P. J. LIOY, G. D. WIGHT, R. E. MEYERS & R. T. CEDERWALL. 1977. An Investigation of Long-Range Transport of Ozone Across the Midwestern and Northeastern U.S. Atmos. Environ. 11: 797.
16. HUSAR, R. B., D. E. PATTERSON, C. C. PALEY & N. V. GILLANI. 1977. Ozone in

Hazy Air Masses. Paper No. 6–5, *In* Proceedings of the International Conference Photochemical Pollution, U.S. EPA Pub. No. EPA-600/3–7–001b, Research Triangle Park, N.C.

17. WEISS, R. E., A. P. WAGGONER, R. J. CHARLSON & N. C. AHLQUIST. 1977. Sulfate Aerosol: Its Geographical Extent in the Midwestern and Southern U.S., Science **195:** 979.

18. WOLFF, G. T., P. J. LIOY & G. D. WIGHT. An Overview of the Current Ozone Problem in the Northeastern and Midwestern U.S., *In* Proceedings of the Mid-Atlantic States Air Pollut. Control Assoc., Conf. on Hydrocarbon Controls, New York, N.Y., April 1977, p. 98.

19. SISTERSON, D. L. & P. FRENZEN. 1978. Nocturnal boundary-layer wind maximum and the problem of wind power assessment. Environ. Sci. Technol. **12**(2): 218.

20. MEYERS, R. Sulfate modeling studies. Presented at the Air Pollution Control Association Conference on the Question of Sulfates. Philadelphia, PA, April 13–14, 1978.

CHARACTERIZATION OF AEROSOLS UPWIND OF NEW YORK CITY: II. AEROSOL COMPOSITION *

Paul J. Lioy † and George T. Wolff †

Interstate Sanitation Commission
New York, New York 10019

Kenneth A. Rahn

Graduate School of Oceanography
University of Rhode Island
Kingston, Rhode Island 02881

David M. Bernstein † and Michael T. Kleinman †

Institute of Environmental Medicine
New York University Medical Center
New York, New York 10016

INTRODUCTION

Airborne particulate matter was collected at rural High Point, New Jersey, during July and August 1976, as part of the New York Summer Aerosol Study (NYSAS). The effort involved the evaluation of the aerosol at this rural site, which is located approximately 55 miles west-northwest of New York City, and the examination of the upwind relationship between the High Point data and data obtained coincidently at the New York University Medical Center, New York City. The information in this paper characterizes the particulate matter collected throughout the period at High Point both chemically and physically. The general sources of the inorganic portions of the particulates are indicated by the results from analysis of the aerosol during periods of elevated sulfate (SO_4) and ozone (O_3) concentrations, and high trace-element enrichment factors.

METHODOLOGY

The aerosol at High Point was characterized using a variety of sampling and analytic techniques. The sampling and continuous monitoring equipment

* Revised version of paper presented at the American Industrial Hygiene Association 1977 Conference, New Orleans, LA, May 22–27, 1977, as part of the New York Summer Aerosol Study (NYSAS).

† Current affiliations:

David M. Bernstein
Brookhaven National Laboratory
Upton, New York 11973

Michael T. Kleinman
Rancho Los Amigos Hospital
University of Southern California
Downey, California 90242

Paul J. Lioy
Institute of Environmental Medicine
New York University Medical Center
New York, New York 10016

George T. Wolff
Environmental and Science Department
General Motors Research Laboratories
Warren, Michigan 48090.

73

was located on a ridge with an elevation of 1600 ft at the New Jersey State Park. Under the influence of the prevailing westerly winds, there are few major source areas upwind of High Point. Therefore, most of the time this site provides an opportunity to examine the general summertime pollution levels upwind of New York City.

Coincident sampling for the NYSAS at High Point and New York City was conducted from August 1 through September 6, 1976. Supplemental data were provided from measurements made at High Point in July 1976.

The equipment operated and procedures employed during the study are described below.

1. Hi-Volume air samplers were used to collect suspended particulates (TSP) for total mass, $SO_4^=$, and trace metals analyses. The samples were collected using two different formats. The first involved daily samples collected on glass-fiber (Gelman-Spectrograde) filters from noon to noon during week (experiment) #1. Then during the subsequent week (experiment #2) two 3½ day samples were collected on Whatman filters. This procedure was then repeated alternately on the third and fourth week of August, 1976.

The filters from weeks (experiments) 1 and 3 were analyzed by the N.Y.U. Institute for Environmental Medicine for total suspended particulates (TSP), SO_4,[1] and heavy metals (atomic absorption). The filters from weeks (experiments) 2 and 4 were analyzed for trace elements at the University of Rhode Island using neutron activation analysis.[2]

2. A standard Royco optical sensor of the forward scattering type was used to determine the aerosol volume of particles between 0.5 μm and 5 μm in diameter. The instrument was operated during July 1976.

3. The light-scattering extinction coefficient, b_{scat}, was measured throughout the study with a standard MRI integrating nephelometer.

4. Ozone data was collected continuously with a Bendix chemiluminescent monitor from July through September.

RESULTS

Particulate Mass Analyses

Experiments 1 and 3

The total suspended particulates (TSP) concentrations detected at High Point had a wide range in values (FIGURE 1), with a maximum concentration of 142 μg/m³ and a minimum concentration of 13 μg/m³.

During the study, a variety of summertime meteorologic conditions influenced High Point. However, during most of the sampling, the weather was dominated by high pressure systems passing through the area. In the cases of weeks (experiments) 3 and 4, the same stagnating high pressure system prevailed over the region. Each of the four separate sampling periods is indicated on FIGURE 1 for point of reference.

In conjunction with the TSP analysis for experiments 1 and 3, water soluble $SO_4^=$ was determined on all filters. The results shown in FIGURE 2 indicate a gradual increase and an eventual decrease in the daily concentration during experiment 1 and significant daily increases of the $SO_4^=$ during experiment 3.

As seen in previous studies,[3-6] the trends in the TSP and $SO_4^=$ concentrations are similar to those observed for the ozone maxima as seen in FIGURES 1 and 2. Simultaneous increases and eventual decreases in TSP, $SO_4^=$, and O_3 have been found to be associated with the movement of anticyclonic weather systems. The peak concentrations appear to occur on the back side or return flow of the anticyclone.

Calculation of the $SO_4^=$/TSP ratio also shows a wide range of variability with a maximum of 0.59 and a minimum of 0.03. The ratio does not necessarily follow the trend for increasing or decreasing sulfates or TSP. For example, on August 23, the study maximum of 27 $\mu g/m^3$ $SO_4^=$ was recorded, and the ratio was 0.30.

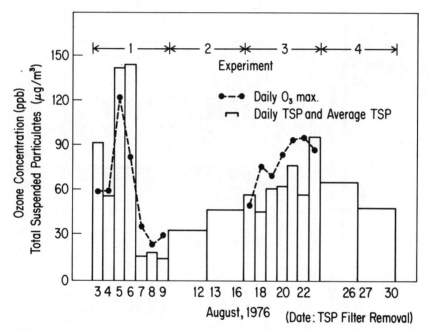

FIGURE 1. Total suspended particulate concentration and daily ozone maximum detected at High Point, N.J. during August, 1976.

Trace Elements

For the same samples, atomic absorption analysis (AAS) was used to determine daily trace metals concentrations and the results are found in TABLE 1. The averages for High Point during experiment 1 were generally 7 to 15 times lower than the trace-metal concentrations found at the New York City site (TABLE 2). However, on August 4 and 5, the Cd and Zn values detected at High Point were >2 times higher than those recorded in New York City. The concentration patterns for Pb, Fe, Zn, Mn, Cr, and Ni appear to follow the trend observed for TSP. A linear correlation of 0.755 was found between TSP and Zn in these experiments, but there were no correlations

TABLE 1

TOTAL SUSPENDED PARTICULATES AND METALS CONCENTRATIONS FOR WEEKS 1 AND 3 OF NYSAS AT HIGH POINT, NEW JERSEY

Date Sampling From 1200N–1200N	TSP ($\mu g/m^3$)	Pb	Fe	Mn	Zn	Cd	Cu	Cr	Ni	V
				(Air Concentrations in $\mu g/m^3$)						
8/2–8/3	91.4	.078	.172	.0034	<.002	<.0005	.024	.0030	.0070	√
8/3–8/4	54.1	.084	.218	.0040	<.002	<.0040	.056	.0020	.0018	√
8/4–8/5	140.1	.130	.378	.0076	.660	.0078	.008	.0016	.0050	√
8/5–8/6	142.8	.182	.420	.0104	.400	.0056	.028	.0004	.0044	√
8/6–8/7	14.0	.090	.120	.0016	.090	<.0005	.010	.0026	.0020	√
8/7–8/8	15.8	.066	.050	.0002	.028	<.0005	.022	.0004	.0022	√
8/8–8/9	13.0	.072	.076	.0008	.002	.0120	.008	.0010	.0024	.0028
8/16–8/17	54.0	.034	.050	<.0004	<.002	<.0005	.028	.0050	.0052	√
8/17–8/18	43.5	.050	.122	.0004	.005	.0040	.002	.0028	.0072	√
8/18–8/19	61.4	.110	.286	.0060	.028	<.0005	.004	.0040	.0040	.032
8/19–8/20	61.6	.174	.372	.0060	.064	<.0005	.050	.0020	.0060	.016
8/20–8/21	75.1	.210	.332	.0066	.144	.0044	.015	.0041	.0068	√
8/21–8/22	56.7	.266	.594	.0144	.066	<.0005	.010	.0048	.0074	√
8/22–8/23	94.6	.280	.660	.0140	.068	<.0005	.058	.0052	.0078	√
Average	65.5	.130	.276	<.0054	<.112	<.0032	.023	.0026	.0050	√

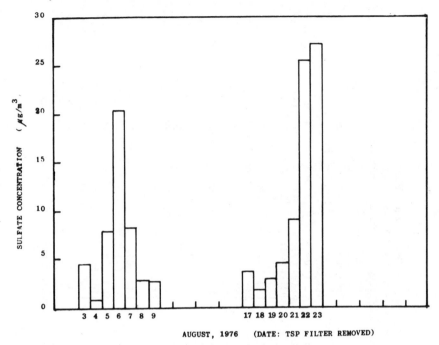

FIGURE 2. Sulfate concentrations detected at High Point, N.J. during August, 1976. Experiments 1 and 3 of the NYSAS.

coefficients above 0.60 for TSP and the other metals. However, Pb had an 0.948 and 0.936 correlation coefficient with Fe and Mn, respectively.

Experiments 2 and 4

In experiments 2 and 4, 3.5 day integrated mass samples were collected at High Point. These samples were analyzed for trace metals using neutron activation (INAA) analysis and the results are reported in TABLE 3. For comparable

TABLE 2

AVERAGE RATIOS OF NEW YORK CITY TO HIGH POINT OF SOME TRACE METALS FOR THE PERIOD AUGUST 2 THROUGH AUGUST 9, 1976

Metal *	NYC/HP Ratio
Pb	12.7
Fe	15.7
Mn	13.6
Cr	15.6
Ni	7.08

* Zn and Cd had a number of concentrations below the detection limit.

TABLE 3

METALS CONCENTRATIONS AND ENRICHMENT FACTORS (EF) AT HIGH POINT, NEW JERSEY FOR EXPERIMENTS 2 AND 4 OF THE NYSAS

Dates	8/9–8/12		8/13–8/16		8/23–8/25		8/26–8/30	
Material	Concentration *	EF	Concentration	EF	Concentration	EF	Concentration	EF
TSP	31.4		44.3		64.8		46.9	
Ba	2.6 (1.0)	2.27	6.6 (0.6)	2.00	8.2 (1.0)	2.50	4.4 (0.7)	2.20
Ti	12.2 (5.0)	1.03	29.5 (5.0)	1.22	63.0 (6.0)	1.69	36.0 (4.0)	1.75
In	.054 (.003)	200	.143 (.005)	260	.021 (.002)	25	.068 (.003)	145
Br	15.9 (.4)	2410	13.4 (.40)	1015	17.7 (.5)	855	15.6 (.5)	1340
Mn	8.5 (.3)	3.3	14.0 (.4)	2.72	15.1 (.4)	1.87	12.0 (.3)	2.70
Mg	70 (30)	1.24	120 (20)	1.06	220 (30)	1.24	150 (30)	1.54
Na	217 (5)	2.83	142 (5)	.93	160 (6)	.67	129 (5)	.98
V	9.0 (.3)	25	7.9 (.3)	10.8	13.4 (.3)	11.7	11.9 (.3)	18.9
Cl	<15	<42.6	<15	<21.3	<15	<13.6	<15	<24.7
Al	220 (5)	1.00	440 (10)	1.00	690 (20)	1.00	380 (10)	1.00
Sm	.031 (.001)	1.90	.062 (.003)	1.91	.130 (.002)	2.54	.051 (.001)	1.82
Au	<.02	<1900	<.006	<278	.0085 (.0021)	250	.0027 (.0017)	145
Zn	90 (5)	480	310 (10)	820	122 (3)	210	192 (3)	590
As	1.40 (.04)	289	2.7 (.04)	279	2.4 (.10)	158	2.6 (.1)	310
Cd	6.6 (1.5)	12200	7.9 (1.3)	7300	<4	<2400	3.3 (.9)	3500
Sb	.56 (.02)	1017	.64 (.02)	581	1.47 (.03)	870	.80 (.03)	860
Ga	.24 (.05)	6.0	.44 (.05)	5.56	.56 (.07)	4.5	.27 (.04)	3.95
Eu	.0054 (.0025)	1.64	.013 (.002)	2.00	.024 (.004)	2.31	.011 (.0015)	1.96
K	148 (6)	2.11	220 (10)	1.56	410 (20)	1.86	210 (10)	1.73
Mo	.80 (.15)	197	1.05 (.12)	129	1.14 (.19)	89.6	.76 (.10)	108
Fe	230 (10)	1.7	385 (15)	1.42	650 (15)	1.53	385 (8)	1.65
La	.23 (.02)	2.82	.46 (.02)	2.80	1.18 (.05)	4.62	.46 (.03)	3.30

* All concentrations in $\mu g/m^3$, except TSP in $\mu g/m^3$.

elements, the concentrations detected in experiments 2 and 4 (INAA) were somewhat higher than those observed in experiments 1 and 3 (AAS).

In addition to the actual concentrations recorded, aerosol-crust enrichment factors have been calculated. Each enrichment factor (EF) was calculated using the equation

$$EF_x = (X/Al)_a / (X/Al)_c$$

where X is a particular element and the subscripts a and c refer to the aerosol and the crust, respectively. The enrichment factors are presented to compare the relative proportion of the elements in the aerosol to the average relative proportions of the same elements in the earth's crust. For example, an enrichment factor of 1.0 indicates the element is in its crustal proportion. Enrichment factors significantly different from 1.0 indicate possible anthropogenic enrichment. The elemental concentrations reported for average crustal rock by Mason [8] were used in the present analysis to be consistent with previous calculations by Rahn,[9] with Al being used as the reference element. The calculated enrichment factors are found in TABLE 3. According to the curves developed by Rahn,[9] the elements In, V, Au, Zn, As, Cd, Sb, and Mo are highly enriched for a rural site. Significant variation occurred in the enrichment factors for these elements. A typical example is Zn, which had enrichment factors of 480 and 820 for experiment 2 (Aug. 9–16), and 210 and 590 for experiment 4 (Aug. 23–30). These higher enrichments generally occurred on west and southwest winds.

The Na enrichments were usually about 1, indicating that the air did not have a maritime origin. However, for the period August 9 through 12, during experiment 2, an enrichment of 2.83 was calculated demonstrating that some maritime air had influenced the area. This could be traced back to August 9 when Hurricane Belle passed over the Northeast. The values for the remaining elements in TABLE 3 are close to unity, which indicates that they are present in soil-like ratios in the aerosol.

Aerosol Volume

Calculations were made for five particle diameter ranges above 0.5 μm to determine the aerosol volume distribution, and the results are shown in FIGURE 3. The results indicate that a bimodal volume distribution exists at High Point which is consistent with typical particle size distributions.[10] The data also demonstrate that much of the volume is in the accumulation size range (0.1 to 2 μm diameter) with a few particles in the coarse particle size range (>2 μm). Further evidence that much of the volume is in the accumulation mode can be derived from trace metal and ozone data obtained on July 21. On this day High Point recorded the largest daily average volume in the 0.5 to 0.7 μm diameter range for the sampling period. In addition, the enrichment calculations show EFs of 330, 111, 230, 1110 for the trace elements of In (Indium), Mo, As, and Zn, respectively, indicating an influence of industrial or urban submicron aerosols. High sulfate concentrations, which are predominately in the submicron size, could also be expected since an ozone maximum of 132 ppb and TSP concentration of 55 μg/m^3 were recorded.

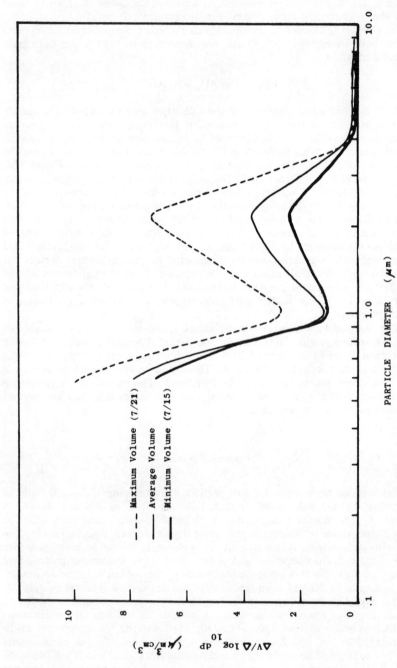

FIGURE 3. The aerosol volume distribution detected at High Point, N.J. during the period July 12–22, 1976, using a forward scattering optical particle counter.

Visibility

As summarized by Miller *et al.*[11] b_{scat} is directly related to visibility. The b_{scat} coefficient was measured during various periods throughout the study. Representative of this data is the period from 16 through 24 August, which includes the third week of sampling (experiment 3). As can be seen in FIGURE 4, the visibility deteriorates during the period. When compared to the O_3, TSP, and $SO_4^=$ levels presented in FIGURES 1 and 2, the visibility deteriorates as the concentrations of these parameters increase.

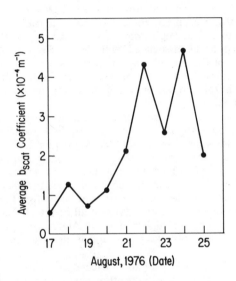

FIGURE 4. b_{scat} data collected at High Point, N.J. August 16–24, 1976.

DISCUSSION

The general variation of the TSP, $SO_4^=$, and O_3 concentrations observed at High Point, as shown in previous studies [3, 5, 12, 13] and Part I of this study,[14] is associated with the presence of high pressure systems (anticyclones) and the greatest accumulation occurs on the return flow or back side of the system. These systems typically spend 4 to 15 days (av = 7) in the northeastern quadrant of the U.S.[15] The simultaneous increase in sulfates and ozone in the summer appear to be a result of the accumulation and oxidation of precursors within the photochemical smog complex and is consistent with the modeling results of Sander and Seinfeld[16] and Graedel *et al.*[17]

In conjunction with the buildup of these secondary pollutants, a reduction in visibility was observed. This is consistent with the phenomena discussed by Burton *et al.*,[18] using the SAI urban air shed model where a coincident increase in particulates and a reduction in visibility is a function of the chemical processes in photochemical smog. The behavior of a significant portion of the aerosol and the visibility in the model appears to be correlated to the concentration of ozone and may be controlled by the same processes.

The trace metal concentrations at High Point demonstrate a periodicity, and the values generally did not correlate well with TSP. However, the Pb, Mn, and Fe did show a high degree of intercorrelation in experiments 1 and 3, indicating the possibility of common source areas. The values for Mn, Pb, and Fe were usually 2 to 4 times lower than those found in New York City.

According to the enrichment factors calculated for experiments 2 and 4, the aerosol is probably influenced by upwind industrial and urban pollution. The high enrichments and concentrations of trace metals like In, As, Zn, V, Cd, and Mo were unusual for a rural site like High Point. A typical example is In, which usually has an enrichment of 30 in a rural area but had values ranging from 25 to 260 at High Point. The occurrence of the highly enriched elements appeared to be associated with W to SW winds, which persisted greater than 50% of the time during experiments 2 and 4. Under such conditions, High Point was downwind of the northeast industrial corridor. For example, the high In, and Zn enrichments would indicate the possibility of influence from smelters.

The trace metal data for experiments 2 and 4 were compared to those obtained for July and August 1970 at Algonquin Provincial Park, Canada,[19] and the results are shown in TABLE 4. Algonquin has been known to be influenced by an industrial area with a large smelter complex, approximately 200 miles to the west,[19] and High Point at times is downwind of major industrial sources. From TABLE 4, it can be seen that the enrichments are similar for nearly all of the detected elements reported at both High Point and Algonquin (within $2\times$ for 13 of 17 elements). Almost all of the elemental concentrations observed at High Point are higher than at Algonquin. Arsenic, which is a smelter emission, is the only exception, and it also had the lowest enrichment factor ratio.

According to Rahn,[9] many of the enriched elements can be classified geochemically as chalcophilic, a classification based on their affinity for mineral sulfide compounds. Most of the elements with this classification are associated with particles having diameters in the coagulation and condensation range, and probably result from high temperature condensation processes. On the basis of this information, and the Algonquin comparison, the high enrichments are thought to be a consequence of industrial emissions. The results suggest that under certain conditions the emissions from the industrial areas in the northeast corridor impact upon High Point. This phenomenon has been documented for ozone within a high pressure system by Wolff et al.,[20] and on the basis of the simultaneous build up of O_3, $SO_4^=$, TSP, and trace metals seen in this study and the results of the transport analysis,[14] the elevated concentration of anthropogenic primary and secondary particles would be expected for west-southwest winds. This effect is supported by two other observations. The detected aerosol volume distribution had a relatively constant shape with a significant portion of the volume in the accumulation size range. In addition, $SO_4^=$,[21-23] and some of the trace metals [9, 24] are present in the submicron size range.

The nonenriched elements appear to have enrichment factors similar to those observed at sites all around the world.[9] The values indicate that the origin of Ti, Mg, Sm, Eu, K, and La could be the earth's crust, although anthropogenic input in soil-like ratios cannot be discounted.

Conclusions

The investigation has demonstrated that high ambient concentrations of TSP and $SO_4^=$ occur at High Point even though its physical location would classify it as a rural site. This is consistent with the observations reported in 1975.[5] The trace-element concentrations at High Point, however, are usually lower than observed in urban areas.

It appears that the secondary pollutants, O_3 and $SO_4^=$, accumulate under the influence of anticyclonic activity. Using the calculated enrichment factors,

Table 4

Comparison of the Average Concentrations and Enrichments for Two Continental North American sites: High Point, New Jersey (rural) and Algonquin Provincial Park, Canada (remote), under Summertime Conditions

	Concentrations			Enrichment Factors		
Elements	High Point	Algonquin	HP/Alg	High Point	Algonquin	Ratio
	(ng/m^3)		Ratio			HP/Alg
Ti	35.0	15	2.33	1.42	1.15	1.23
Br	15.7	5.7	2.75	1412	774	1.82
Mn	12.4	12	1.03	2.64	4.82	.54
Mg	140	40	3.50	1.27	.62	2.04
Cu	12.7	7.9	1.61	40.0	48.6	.82
Na	162	69	2.35	1.35	.83	1.63
V	10.6	1.9	5.74	16.6	4.75	3.49
Cl	<15	<4	<5.20	<25.5	<10.4	<2.44
Sm	.068	.051	1.33	2.01	2.89	.69
Zn	179	40	4.47	525	194	2.70
As	2.27	4.7	.48	259	884	.29
Sb	.87	.60	1.45	832	1020	.82
Ga	.38	.14	2.71	5.0	3.14	1.59
Eu	.0134	.009	1.49	1.98	2.57	.77
K	247	170	1.45	1.82	2.22	.82
La	.59	.30	1.97	3.47	3.39	1.02
Al	433	240	1.80	1.00	1.00	1.00

we see that some of the trace metal constituents also accumulate in the air mass and appear to have an industrial and/or urban origin.

The association of high concentrations of $SO_4^=$ and TSP, reduced visibility, and high concentrations of O_3 demonstrate that High Point is at times influenced by particulate pollutants of a nonlocal anthropogenic origin. Since a significant portion of these particles are in the accumulation size range, long-range transport (on the order of hundreds of miles) influences High Point (see Part I of this study).

ACKNOWLEDGMENTS

The authors wish to thank Dr. Ralph Pasceri, New Jersey Department of Environmental Protection, for the use of some of the sampling equipment; Mr. William Bailey of High Point State Park for providing the facilities; and Messrs. Konrad Wisniewski and Edward Edwards, Interstate Sanitation Commission, for the operation of the High Point site.

REFERENCES

1. KNEIP, T. J., B. P. LEADERER, D. M. BERNSTEIN & G. T. WOLFF. 1979. The New York Summer Aerosol Study (NYSAS) 1976. Am. N.Y. Acad. Sci. **322**. (This volume).
2. DAMS, R., J. A. ROBBINS, K. A. RAHN & J. W. WINCHESTER. 1970. Anal. Chem. **42:** 861.
3. STASIUK, W. N., P. E. COFFEY & R. F. MCDERMOTT. 1975. Relationships between Suspended Particulates and Ozone at a Non-Urban Site. APCA Paper 75–62.7, In Proceedings of the 68th Annual Meeting of the Air Pollution Control Assoc. Boston, Mass., 1975.
4. HIDY, G. M. 1975. Summary of California Aerosol Characterization Experiment. J. Air Pollut. Control Assoc. **25:** 1106.
5. LIOY, P. J., G. T. WOLFF, J. S. CZACHOR, P. E. COFFEY, W. N. STASIUK & D. ROMANO. 1977. Evidence of High Atmospheric Concentrations of Sulfates Detected at Rural Sites in the Northeast. J. Environ. Sci. Health, Part A **12:** 1.
6. WOLFF, G. T. & P. J. LIOY. 1977. Transport of Suspended Particulates into the New York Metropolitan Area. In Proceedings of the 70th Annual Meeting of the Air Pollution Control Assoc., Toronto, Ontario, June 1977.
7. KNEIP, T., D. M. BERNSTEIN & M. T. KLEINMAN. 1979. Data from the New York Summer Aerosol Study, 1976. Personal communication.
8. MASON, B. 1966. Principles of Geochemistry. 3rd edit. pp. 44–46. John Wiley & Sons. N.Y.
9. RAHN, K. A. 1976. The Chemical Composition of the Atmospheric Aerosol. Technical Report Graduate School of Oceanography, University of Rhode Island, July 1976.
10. WILLEKE, K. & K. T. WHITBY. 1974. Physical Characteristics of Denver-Area Aerosols. APCA Paper 74–262, In Proceedings of the 67th Annual Meeting of the Air Pollution Control Assoc., Denver, Colo., June 1974.
11. MILLER, D. F., W. E. SCHWARTZ, J. L. GEMMA & A. LEVY. 1975. Haze Formation: Its Nature and Origin—1975. U.S. EPA Pub. No. EPA 650/3–75–010, Research Triangle Park, N.C., March 1975.
12. MELO, O. T. 1976. Sulfate in Ontario Air. Report #76–322–K, Ontario Hydro Research Division, July 29, 1976.
13. HIDY, G. M., E. Y. TONG, P. K. MUELLER, S. RAO, I. THOMSON, F. BERLUNDI, D. I. MULDOON, D. MCNAUGHTEN, MAJAHAD. 1976. Design of the Sulfate Regional Experiment (SURE). EPRI Publication, February 1976, EPRI–EC 125.
14. WOLFF, G. T., P. J. LIOY, B. D. LEADERER, M. T. KLEINMAN & D. M. BERNSTEIN. 1979. Characterization of Aerosols Upwind of New York City, I. Transport. Ann. N.Y. Acad. Sci. **322**. (This volume.)
15. WIGHT, G. D., G. T. WOLFF, P. J. LIOY, R. E. MEYERS & R. T. CEDERWALL. 1977. Formation and Transport of Ozone in the Northeast Quadrant of the U.S. In Proceedings of ASTM Conference on Air Quality Meteorology and Atmospheric Ozone, Boulder, Colo., Aug. 1977.

16. SANDER, S. P. & J. H. SEINFELD. 1976. Chemical Kinetics of Homogeneous Atmospheric Oxidation of Sulfur Dioxide. Environ. Sci. Technol. **10:** 1114.

17. GRAEDEL, T. E., L. A. FARROW & T. A. WEBER. 1976. Kinetics Studies of the Photochemistry of the Urban Atmosphere Atmos. Environ. **12:** 1095.

18. BURTON, C. S., T. N. JERSKEY & S. D. REYNOLDS. 1977. A Preliminary Investigation of Expected Visibility Improvements in Los Angeles Basin for Oxidant Precursor Gases and Particulate Emission Controls. *In* Proceedings of the International Conference on Photochemical Oxidant Pollution and its Control: Vol. II, EPA 600/3–77–001b, January, 1977.

19. RAHN, K. A. 1971. Trace Elements in Aerosols: An Approach to Clean Air, Ph.D. Thesis, Universtiy of Michigan, 1971.

20. WOLFF, G. T., P. J. LIOY, R. E. MEYERS, R. CEDERWALL, G. D. WIGHT, R. S. TAYLOR & R. E. PASCERI. 1977. Anatomy of Two Ozone Transport Episodes in the Washington, D.C. to Boston Corridor. Environ. Sci. Technol. **11:** 506.

21. Position Paper on Regulation of Atmospheric Sulfates. Strategies and Air Standards Division, U.S. EPA, Research Triangle Park, North Carolina, September 1975.

22. CUNNINGHAM, P. E. & S. A. JOHNSON. 1976. Spectroscopic Observations of Acid Sulfate in Atmospheric Particulate Samples. Science **191:** 77.

23. BERNSTEIN, D. M. & K. RAHN. 1979. New York Summer Aerosol Study: Trace Element Concentration vs. Particle Size. Ann. N.Y. Acad. Sci. **322.** (This volume.)

24. MUELLER, P. K., R. W. MOSLEY & L. B. PIERCE. 1972. Chemical Composition of Pasadena Aerosol by Particle Size and Time of Day IV. Carbonate and Noncarbonate Carbon Content. *In* Aerosols and Atmospheric Chemistry. G. M. Hidy, Ed. Academic Press. New York.

NEW YORK SUMMER AEROSOL STUDY: TRACE ELEMENT CONCENTRATIONS AS A FUNCTION OF PARTICLE SIZE

David M. Bernstein *

*Institute of Environmental Medicine
New York University Medical Center
New York, New York 10016*

Kenneth A. Rahn

*Graduate School of Oceanography
University of Rhode Island
Kingston, Rhode Island 02881*

INTRODUCTION

The size distributions and elemental compositions of the urban atmospheres are important parameters both for characterizing the aerosol itself and for understanding its relationship to sources and meteorology. In addition, these factors are essential for defining the composition of the fractions of the aerosol associated with health effects.

The volume or mass distribution of the urban aerosol is often described by two or three log-normal modes. In New York City the two prevailing modes are the accumulation mode (mass-median aerodynamic diameter (MMAD) between approximately 0.1 and 2.5 μm) and the coarse-particle mode (MMAD greater than approximately 2.5 μm).[1] These two modes are generally attributed to different types of particle production.[2] In general, the fine-particle mode results from vapor phase to particle transformation, whereas the coarse-particle mode results from comminution processes. There is apparently little mixing between the two modes.

To determine the elemental composition of the New York aerosol as a function of size, during the New York Summer Aerosol Study (NYSAS) in August 1976, three different types of size-selective samplers and one high-volume filter sampler were operated. The data obtained were used to achieve a better understanding of the sources and health hazard potential of the New York aerosol.

EXPERIMENTAL

A cyclone size-selective sampler [4] was used to obtain measurements of the size distribution of trace elements at New York University Medical Center (New York, NY) since March, 1976. This device uses parallel two-stage

* Current affiliation:
        Medical Department
        Brookhaven National Laboratory
        Upton, New York 11973

87

0077-8923/79/0322-0087 $01.75/0 © 1979, NYAS

samplers, i.e., cyclone precollectors followed by high-efficiency filters that collect the smaller particles which pass through the cyclones. A parallel array of four of these two-stage samplers plus one total-suspended-particulate sampler, all operating with the same inlet velocity, allowed elemental concentrations to be determined in the following size regions: <0.5 μm, 0.5–1.5 μm, 1.5–2.5 μm, 2.5–3.5 μm, and >3.5 μm MMAD. One-week sampling periods provided four weekly-average size-distribution measurements during the NYSAS program.

A 28.3 ℓ min^{-1} (1 cfm) Andersen 2000, Inc. (Atlanta, GA) cascade impactor was operated for periods of 5 days during the second and fourth weeks of the study. The sampler is a multi-stage, multi-jet cascade impactor. Seven stages and a backup filter provided size ranges of >11.9 μm, 11.9–7.1 μm, 7.1–4.3 μm, 4.3–2.7 μm, 2.7–1.42 μm, 1.42–0.84 μm, 0.84–0.55 μm, and <0.55 μm MMAD. Collection surfaces of uncoated polyethylene were used to facilitate analysis by instrumental neutron activation analysis.

A 0.57 m^3 min^{-1} (20 cfm) Sierra (Carmel Valley, CA) 5-stage, multi-slot cascade impactor was also operated for the second and fourth weeks of the NYSAS program. The stages together with a backup filter provided particle size ranges of >4 μm, 2–4 μm, 1–2 μm, 0.5–1 μm, 0.25–0.5 μm and <0.25 μm MMAD. Whatman No. 41 cellulose filters were used as collection surfaces; Delbag Microsorban was used for the backup filters.

The aerosol samples were analyzed by atomic absorption spectrophotometry (AAS) at New York University (New York, NY) and by instrumental neutron activation analysis (INAA) at the University of Rhode Island (Kingston, RI).

RESULTS

Comparison of Samplers

The schedule of sampling during the second week of August permitted comparison of the total concentrations of trace elements collected by the four samplers (TABLE 1). The results concerning the loss of particles in impactors

TABLE 1

SAMPLER COMPARISON
(August 10–16, 1976)

Sampler	Concentration, ng/m³					
	Pb	Fe	Mn	Cd	Cu	V
Cyclone (AAS)	1120	1520	26	6	56 †	60
Anderson (INAA)	—	1447	34	—	75 †	46
Sierra (AAS)	975	1160	54	28	1550 †	64
* Hi-Vol (INAA+AAS)	790	1547	40	5	858 †	74

AAS—Atomic Absorption Spectrophotometry.
INAA—Instrumental Neutron Activation Analysis.
  * 5-Day average.
  † These numbers reflect the copper contamination as discussed in the text.

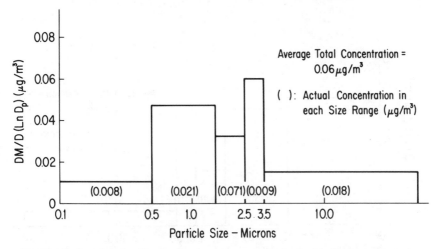

FIGURE 1. Average size distribution for copper as obtained with the cyclone sampler during August 1976 showing the contamination from the hi-vol blowers (see text).

confirmed those reported by others.[5, 6] While the general agreement between samplers was good, severe copper contamination was evident, and was attributed to the pumps used with the Sierra and other hi-vol samplers. Although the cyclone sampler was separated laterally from the hi-vol samplers by more than 4 meters, its copper size distribution was also affected by the pumps. FIGURE 1 shows the quasi-unimodal copper distribution obtained with the cyclone sampler, which differs significantly from the bimodal distribution and the lower concentrations normally found in New York City.[1]

Neither wall losses nor smearing from stage to stage were observed with the Andersen sampler, but both were clearly noticeable in the Sierra. Because of the sampling bias in the Sierra instrument, data from only the cyclone and Andersen samplers were used in further interpretations.[11]

Elemental Size Distributions

The average size distributions of lead, iron, and manganese for August, 1976 are shown in FIGURES 2–4. These distributions are typical of the various types normally encountered in New York City. Lead was found predominantly (80%) in the fine-particle mode (FIGURE 2). This agrees with the publications of many authors, which indicate that the fine particle distribution of lead occurs because the combustion of leaded gasoline is the primary lead source.

In contrast, iron is primarily associated (80%) with coarse particles (FIGURE 3). While many types of comminution sources may produce iron-containing particles, corrosion of automobiles and building materials are probably two of the major anthropogenic sources. A large natural and anthropogenic soil component of iron is also thought to be present in the urban aerosol.

The size distribution for manganese is bimodal (FIGURE 4). This type of distribution indicates two sources, neither of which are known at the present.

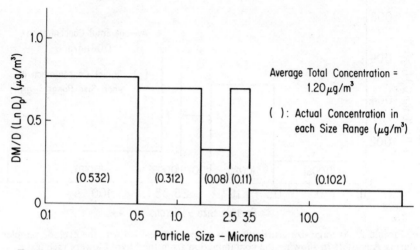

FIGURE 2. Average size distribution for lead as obtained with the cyclone sampler during August 1976.

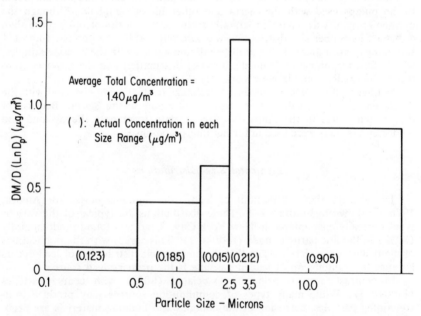

FIGURE 3. Average size distribution for iron as obtained with the cyclone sampler during August 1976.

Coarse-particle manganese may have sources similar to iron; fine-particle manganese probably has a variety of sources, one of which may be the manganese additive to gasoline which was introduced as a substitute for lead additives.

A total of 37 elements were determined by both AAS and INAA. For each element TABLE 2 lists its typical dominant mode(s), the percent of its mass associated with particle sizes less than 2.5 μm MMAD, its total concentration, and potential sources.[7-9] The elements are grouped according to total concentrations. There are several very significant bits of information in this table, which describe the general distributions of the elements. Of the elements with concentrations greater than 500 ng/m³, only lead is primarily associated with fine particles. This is of particular importance in light of the toxicity of lead in man and the contribution of inhaled lead to the total body burden of lead.

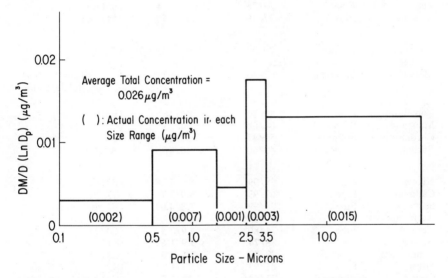

FIGURE 4. Average size distribution for manganese as obtained with the cyclone sampler during August 1976.

Sodium, which was expected to be in the coarse mode from either soil or sea salt sources, was found to be bimodal. The source of the fine particle sodium is not known.

Chlorine, bromine, and iodine can often be used as indicators of the marine aerosol, provided that it is recognized that coarse particle chlorine and bromine and fine-particle iodine must be used. In our study, chlorine was indeed found primarily in the coarse mode but bromine was bimodal. The fine particle bromine was probably from combustion of leaded gasoline. Iodine was essentially undetectable during the first week of sampling with only slightly greater concentrations during the second week, which suggests that it had very limited pollution sources. More detailed information on these halogens in New York City is given in an accompanying paper by Rahn et al.[12]

TABLE 2

CONCENTRATIONS, SIZE DISTRIBUTIONS, AND SOURCES OF TRACE METALS
IN NEW YORK CITY
(August 10–16, 1976/August 23–30, 1976)

	Mode	% <2.5 μm	Total Concentration (ng/m³)	Potential Sources *
>500 ng/m³				
Iron	Coarse	34/28	1447/1136	(C)(A)(S)
Sodium	Bi-modal	61/57	930/1002	(M)(C)
Aluminum	Coarse	19/23	824/949	(C)
Lead	Fine	89/90	1100/1500	(A)
100–500 ng/m³				
Chlorine	Coarse	33/36	350/593	(M)(A)
Potassium	Bi-modal	57/53	313/391	(M)(C)
Magnesium	Coarse	28/18	308/589	(M)(C)
Zinc	Fine	79/78	284/174	(S)(A)
Bromine	Fine	77/79	171/214	(M)(A)
50–100 ng/m³				
Titanium	Coarse	29/23	84/80	(C)
Copper	—	(see text)	—	(I)
Mercury	Bi-modal	67/70	81/62 †	(C) coal combustion crustal degassing
25–50 ng/m³				
Vanadium	Fine	80/70	46/36	(F)
Manganese	Bi-modal	64/68	34/55	(C)(A)(I)
Chromium	Bi-modal	51/45	32/25	(C)(S)
Barium	Bi-modal	49/40	30/46	(A)(C)
1–25 ng/m³				
Nickel	Fine	99/99	18.0/17.8	(F)(S)
Cadmium	Fine	67/99	6.0/3.7	(S)
Selenium	Fine	94/99	7.8/4.0	(F) coal combustion
Antimony	Fine	87/89	6.6/5.8	(S)
Cerium	Fine	71/80	3.9/3.5	(C)(?)
Arsenic	Fine	88/74	3.9/1.6	(S)(C)
Lanthanum	Fine	79/78	3.1/3.0	(C)(?)
Silver	Fine	81/93	2.37/1.24	(S)
Cobalt	Bi-modal	72/44	2.35/3.50	(C)(A)(I)
Molybdenum	Fine	74/—	1.50/—	(S)
<1 ng/m³				
Scandium	Coarse	27/24	0.209/0.244	(C)
Gallium	Bi-modal	80/5 ‡	0.402/0.151	(C)
Indium	Fine	99/99	0.042/0.015	(S)
Iodine	Fine	99/75	0.36–6.7/2.0–2.7	(M)
Cesium	Bi-modal	61/65	0.35/0.12	(M)(C)
Samarium	Fine	77/76	0.283/0.350	(C)(?)
Europium	Bi-modal	72/01 ‡	0.047/0.019	(C)(?)
Lutetium	Coarse	—/—	0.031/—	(C)
Hafnium	Coarse	—/—	0.107/0.152	(C)
Tungsten	Bi-modal	99/01 ‡	0.47/0.329	(?)
Gold	Bi-modal	54/42	0.171/0.199	(?)
Thorium	Coarse	19/15	0.129/0.219	(C)

* (M) Marine; (C) Crustal material, coal fly ash, construction dust; (A) Automotive; (I) Incinerators; (S) Smelting; and (F) Fuel oil, coal fly ash.

† Unusually high value—explanation not yet known.

‡ The modal disparity for these elements stems from their being detectable only on a single stage.

Noncrustal vanadium is a tracer for the combustion of residual fuel oil for heating and power production.[10] Like lead, vanadium stems primarily from combustion and is therefore largely associated with fine particles. The low consumption of fuel oil for heating during this summer period resulted in low vanadium concentrations. New York City in the winter typically has vanadium concentrations about three times greater.[9]

Toxic elements such as cadmium, selenium, antimony, and arsenic were primarily associated with fine particles and were therefore almost completely respirable. Their total concentrations were, however, very low.

The rare-earth elements were generally found at concentrations below 25 ng/m³ and are thought to originate mainly from the crust. It was, therefore, not surprising that 3 out of the 5 measured had major coarse-particle modes. What was unexpected, however, was the important fine particle mode for the lighter rare earth elements La, Ce, Sm, and Eu. This was something of a departure from past measurements,[7] and indicates the presence of a still unknown source.

Total Trace-Element Data

The total concentrations of trace elements measured during August 1976 were examined both in terms of enrichment relative to the earth's crust and their associations with tracers for various sources. Atmospheric aerosols can be broadly divided into pollution-derived and natural particles. The primary natural sources of aerosols collected near the earth's surface would be expected to be the crust and the sea. New York, on the edge of both continent and ocean, should have both crustal and marine components in its aerosol. Rahn [7] has defined an aerosol-crust enrichment factor as follows:

$$EF_x = \frac{(X/Al) \text{ Aerosol}}{(X/Al) \text{ Rock or soil}}$$

where Al is the crustal reference element and average crustal rock is the reference material. Average elemental concentrations and enrichment factors for New York City and other locations are listed in TABLE 3, in descending order of their enrichment factors.

Nonenriched elements such as Al, Fe, Sc, Mn, Ti are present in the New York City aerosol in crustal proportions. While these elements are likely to be associated with soil they may in fact be derived either from soil or from fly ash.

Nearly one-half to two-thirds of the elements have moderate to high enrichment factors. The enrichment factors for Br, Cl, and I indicate that the aerosol in New York City has characteristics of marine sources. Most of the enriched elements, however, are not accounted for by the sea and must have nonmarine and noncrustal origins. While in general these enriched elements are assumed to be associated with pollution sources, similar high enrichment factors are sometimes found in nonpollution areas where natural sources are more important.

Comparing the enrichment factors for New York City with those for other cities and marine aerosols two of the most heavily enriched elements (Pb and Br) are at least partially associated with automotive sources. Antimony, cadmium, and zinc, which are also highly enriched, are often associated with smelting. Although few smelting sources exist in New York City, these elements may be transported on fine particles from areas of major smelting activity well outside New York City.

TABLE 3

AVERAGE CONCENTRATIONS AND ENRICHMENT FACTORS

| Element | Average Concentration N.Y.C. ng/m³ | Enrichment Factors | | | |
		New York City	Philadelphia [7]	Bermuda [7]	Tucson,[7] Arizona
Hg	72	88181 *	680	— †	—
Pb	1200	9867	4200	92	379
Se	7.8	11642	1080	—	—
Br	171	7597	430	—	—
Au	0.19	4563	—	—	—
Sb	6.6	3059	8100	—	—
Ag	1.9	2544	230	—	—
Cd	6.0	2393	136	338	209
Cl	350	358	21	—	—
Zn	284	323	194	15.2	19.4
I	1.2–4.4	235–800	5.4	—	—
As	3.9	151	75	—	—
Mo	1.5	99	36	—	—
V	46	30	40	—	—
Cr	31	28	10.8	0.92	0.49
In	0.03	28	—	—	—
W	0.40	26	18	—	—
Ni	30	24	36	—	1.02
Co	2.9	12	22	—	0.64
La	3.1	10	18	—	—
Ba	48	8.8	2.6	—	—
Lu	0.03	6.1	—	—	—
Ce	3.7	6.1	9.0	—	—
Sm	0.32	5.2	4.5	—	—
Mn	45	4.6	5.7	0.56	0.54
Hf	0.13	4.3	1.8	—	—
Na	966	3.4	0.96	13.6	0.57
Eu	0.03	2.7	4.5	—	—
Fe	1292	2.5	2.2	0.97	0.69
Th	0.17	2.4	—	—	—
Mg	449	2.1	6.5	2.5	0.71
Ti	82	1.8	2.5	—	1.26
Ga	0.28	1.8	11	—	—
K	352	1.3	0.31	1.28	1.19
Sc	0.23	1.0	12.3	—	—
Al	889	1.0	1.0	1.0	1.0
Cs	0.24	0.39	3.6	—	—

* =Unusually high value—explanation not yet known.
† —=No data available.

Pollution-derived elements usually are more enriched in urban environments than in rural areas. Thus the large enrichment of lead in New York City as compared to Tucson, Arizona would be expected. The daily lead concentration during August, and daily traffic flow on the highway adjacent to the sampling site are shown in FIGURE 5. Using the Spearman nonparametric statistic a significant positive association ($p < 0.02$) was found between traffic flow and the concentration of lead in the air during the period of August 10–31. Data from prior to August 10 were not included due to the onset of hurricane Belle.

Meteorologic factors can create similar concentration trends for elements

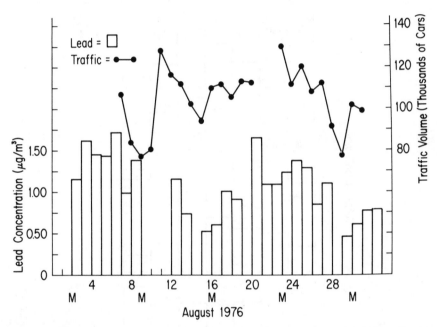

FIGURE 5. Total daily lead concentration vs. traffic volume on a highway adjacent to the sampling site during August 1976. The minimum in traffic flow from August 8 through 10 was due primarily to a hurricane.

that really have very different sources. An example of this is given in FIGURE 6, which shows daily average concentrations for vanadium, cadmium, manganese, iron, and lead through August. Iron, manganese, and cadmium were all found to be significantly correlated with vanadium ($p < 0.02$) and all were found to have the greatest concentrations during the Thursday–Friday periods. This corresponds with weather conditions during August, which exhibited weekly cycles of increasing stability through Thursday, followed by a changing weather system. Lead, however, behaved independently from the other elements in FIGURE 6. The lack of a regular cycle for lead may be the result of the strong local automotive source for lead, but neither traffic data nor other observations bear out such a hypothesis.

CONCLUSION

The trace-element data collected during the NYSAS program have improved our understanding of the sources and size distributions of the aerosol and its constituents in New York City.

The size distributions and enrichment factors of the trace elements in the New York City aerosol indicate the major pollution sources to be gasoline and fuel oil combustion. A significant association was determined between highway traffic volume and the concentration of lead adjacent to the highway. Particu-

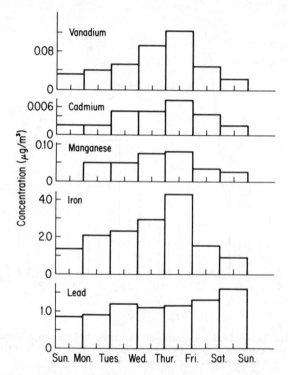

FIGURE 6. Daily average concentrations for vanadium, cadmium, manganese, iron and lead through August 1976.

lates associated with these combustion sources were found to be largely of respirable size. In addition, the data indicate that natural marine and crustal components can also be present in the New York City aerosol. There is also evidence that some material is transported into the New York City area from sources upwind of the city.

SUMMARY

The size distribution and elemental composition of the urban atmosphere are important parameters both for characterizing the aerosol itself and for understanding its relationship to sources and meteorology. The elemental concentrations of 37 trace elements in the New York City aerosol were determined as a function of size, during the New York Summer Aerosol Study in August 1976, using two cascade impactors (20 CFM Sierra and 1 CFM Andersen), and one parallel flow sampler using a cyclone precollector upstream of each filter. Microsorban filters were used for all inline filtering, Whatman 41 on the Sierra impaction stages and polyethylene sheets on the Andersen impaction stages. The 1 CFM Andersen samples were analyzed by Instrumental Neutron Activation Analyses alone. All others were analyzed by Atomic Absorption Spectrophotometry as well. The size distributions and enrichment ratios of trace elements in the New York City aerosol, relative to the elemental composition of the earth's crust, indicate the major pollution sources to be gasoline and fuel oil combustion. A significant association was determined between highway traffic volume and the concentration of lead adjacent to the highway. Particulates associated with these combustion sources were found to be largely of respirable size. In addition, the data indicate that natural marine and crustal components can also be found in the New York City aerosol. Furthermore, some material may be transported into the New York City area from upwind smelting operations.

REFERENCES

1. BERNSTEIN, D. M. 1977. The Influence of Trace Metals in Disperse Aerosols on the Human Body Burdens of Trace Metals. Thesis, New York University. University Microfilm.
2. WHITBY, K. T. 1973. On the Multimodal Nature of Atmospheric Aerosol Size Distributions. Paper presented at the 7th International Conference on Nucleation, Leningrad, U.S.S.R.
3. LIPPMANN, M. 1977. Regional Deposition of Particles in the Human Respiratory Tract. In Handbook of Physiology, Section 9—Reactions to Environmental Agents. D. H. K. Lee, Ed. American Physiological Society. Bethesda, Md.
4. BERNSTEIN, D. M., M. T. KLEINMAN, T. J. KNEIP, T. L. CHAN & M. LIPPMANN. 1976. A High-Volume Sampler for the Determination of Particle Size Distributions in Ambient Air. J. Air Pollut. Control Assoc. 26: 1069–1072.
5. WESOLOWSKI, J. J., ET AL. 1976. Collection Surfaces of Cascade Impactors. In Proceedings of Symposium on X-ray Fluorescence Analysis of Environmental Samples. Research Triangle Park, North Carolina.
6. KNUTH, R. H. 1976. Comments on Inertial Impactor Calibration and Use. Aerosol Measurement Workshop. Gainsville, Florida.
7. RAHN, K. A. 1976. The Chemical Composition of the Atmospheric Aerosol. Technical Report. Graduate School of Oceanography, University of Rhode Island, Kingston, RI.
8. EISENBUD, M. & T. J. KNEIP. 1976. Trace Metals in Urban Aerosols. Electric Power Research Institute Report 117, NTIS Pub. No. Pb-248-324.
9. KLEINMAN, M. T., D. M. BERNSTEIN & T. J. KNEIP. 1977. Seasonal and Source Relationships for Urban Suspended Particulate and Trace Element Concentrations in New York City. J. Air Pollut. Control Assoc. 27: 65–68.
10. KNEIP, T. J., M. EISENBUD, C. D. STREHLOW & P. C. FREUDENTHAL. 1971. Airborne Particulates in New York City. J. Air Pollut. Control Assoc. 20: 144–149.
11. WALSH, P. R., K. A. RAHN & R. A. DUCE. 1978. Env. Sci. and Tech. In press.
12. RAHN, K. A., R. D. BORYS, E. L. BUTLER & R. A. DUCE. Gaseous and particulate halogens in the New York City atmosphere. Ann. N.Y. Acad. Sci. 322. This volume.

CHEMICAL COMPOSITION OF SULFATE AS A FUNCTION OF PARTICLE SIZE IN NEW YORK SUMMER AEROSOL *

Roger L. Tanner, Robert Garber, and William Marlow

Atmospheric Sciences Division
Department of Energy and Environment
Brookhaven National Laboratory
Upton, New York 11973

Brian P. Leaderer

Department of Epidemiology and Public Health
John B. Pierce Foundation Laboratory
Yale University School of Medicine
New Haven, Connecticut 06519

Marie Ann Leyko

New York University Medical Center
New York, New York 10016

INTRODUCTION

In August, 1976, aerosol samples were collected by 12-hour low volume sampling using a sampler based on the Sinclair diffusion battery, by 24-hour averaged, 6 min/hr high volume sampling, by conventional 24-hour high-volume sampling, and by sampling with a high-volume, 5-stage Andersen impactor as a part of the New York Summer Aerosol Study (NYSAS). NYSAS was an intensive, multilaboratory effort to characterize the New York City aerosol during the summer season with respect to number and mass distributions, optical properties, organic content, as well as elemental composition and concentrations of key ionic species such as sulfate and nitrate as a function of aerosol particle size. The sampling site, the array of instrumentation therein and the historical background pertaining to NYSAS are described in detail elsewhere.[1] We describe in this paper the results of the aerosol sampling and chemical analysis for sulfate and related species (specifically including ammonium). In particular, we will concentrate on correlations between aerosol sulfate concentrations and levels of other ionic aerosol constituents and on their variations (or lack thereof) with aerosol particle size.

* Presented at the American Industrial Hygiene Assoc. Meeting, Aerosol Technol. Symposium, New Orleans, Louisiana, May 22–27, 1977.

This work was performed under the auspices of the United States Department of Energy under Contract No. EY–76–C–02–0016.

99

EXPERIMENTAL

Sampling

All aerosol data reported herein were obtained from samples collected at the NYU Medical Center rooftop site near the East River in downtown Manhattan as described elsewhere.[1] The diffusion sampler was located inside the protective shed and sampled air at 47 1/min through the sampling line described by Knutson et al.[3] The BNL Staplex high-volume sampler was located outdoors in a protective, louvered box and filters were changed at 1200 hr each day. This sampler was connected to a timer so that aerosol was sampled during the same 6 minutes of each hour that Knutson et al. were collecting number distribution data by the ATB-CFC apparatus.[3] The 24-hr high-volume sampler and the Andersen impactor (HiVol with impactor head attached) were housed outdoors in conventional aluminum shelters. Both high-volume samplers and the cascade impactor were changed at approximately 1200 hr each day and the data reported for a given day (e.g., day 45, August 25, 1976) is an average value for the time period from 1200 hr that day to 1200 hr the following day.

The low-volume diffusion sampler was operated for 12-hr intervals from about 0730 hr to 1930 and 1930–0730 hr daily corresponding roughly to daylight and darkness hours. During each sampling period three low volume samples were collected: DBU—untreated aerosol as it arrived from the sampling line at 12.2 1/min; DB1—aerosol with 50% cutoff diameter of $\sim$0.035 μm, at 19.9 1/min; and DB2—aerosol with 50% cutoff diameter of $\sim$0.13 μm, at 16.0 1/min. As described elsewhere,[2,4] we stress that the diffusion processor preferentially removes smaller particles because of their higher diffusion coefficient. The 50% cutoff diameter—e.g., $d_{50\%} = 0.035/\mu$m for DB1 samples —signifies that 50% of the particles by numbers of that diameter will penetrate that portion of the battery, 50% will be captured. Particles of smaller diameter will be removed at >50% efficiency, larger particles at <50% efficiency; fractional penetration curves for these sampling conditions are shown in FIGURE 1.

Chemical Analysis

The 24-hr samples from the 5-stage Andersen impactor [5] and the Universal 24-hr high-volume sampler were extracted in 200 mℓ of hot water (approximately 100° C) and portions of the extracts analyzed for sulfate by barium sulfate gravimetry. Additional analyses of sample extracts for nitrate were performed on the Autoanalyzer using the hydrazine-reduction, colorimetric technique [6]; identical analyses for nitrate were also performed on both BNL HiVol and low-volume samples. These data are discussed in a separate paper by Kleinman et al.[7] Some of the impactor and Universal HiVol samples were also analyzed for sulfate by direct-injection enthalpimetry [8]; but only limited data were obtained and these are discussed elsewhere.[9]

Portions of all BNL HiVol and low volume (diffusion sampler) airborne particle samples were extracted in 10 mℓ of room-temperature, 10^{-4} N H_2SO_4 and analyzed for strong acid by the Gran titration method,[10,11] ammonium by indophenol colorimetry,[12] sulfate by $BaCl_2$ turbidimetry,[13] and nitrate as described above. The use of H_3PO_4-treated quartz filter material prevents artifact sulfate formation and allows determination of as little as 0.05μeq of strong acid.[11]   Some particle samples were also analyzed for total sulfur content by the reduction-silver-110 method [14] and specifically for sulfuric acid by the combination of benzaldehyde extraction [11] and flash volatilization-flame photometric detection.[15, 16]

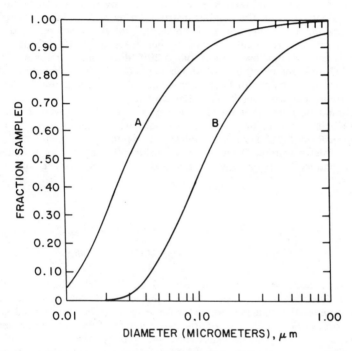

FIGURE 1. Fractional penetration curves for diffusion processor samples during NYSAS, August, 1976; Curve A—sample DB1 collected at 19.9 ℓ/min after 7 sections of diffusion battery; Curve B—sample DB2 collected at 16.0 ℓ/min after all 11 sections of diffusion battery.

Diffusion Sampling Size Distribution Analysis

As reported by Knutson *et al.*,[3] size distribution measurements were made by the diffusion battery-continuous flow counter (ATB-CFC) instrument for 230 hr and by the electrical aerosol analyzer (EAA) for 410 hr during NYSAS. Aerosol size distributions by number were calculated from EAA data by the conventional method and also from the ATB-CFC data as described by Knutson *et al.*[3] The essential step was the application of Twomey's nonlinear iterative algorithm [17] to obtain the aerosol concentration in eight size classes, geometrically distributed from 0.0032 to 0.32 μm diameter. We have also recalculated the EAA data using Twomey's algorithm by application of an 18-particle-size calibration matrix kindly supplied by Dr. Benjamin Liu of the University of Minnesota.

From the product of the fractional penetration curves in FIGURE 1 and the average size distribution by number for a given sampling period calculated from either EAA or ATB-CFC data, one obtains the average number distribution reaching the diffusion-processed samples DBU, DB1, and DB2 collected during that period. The average size distribution by surface and volume may then be calculated from the number distribution by making the usual spherical particle assumption. Comparison of the aerosol volumes reaching the diffusion-processed samples as calculated from either EAA or ATB-CFC data, where simultaneously taken, showed more than a 20% difference in volume in only 30 of the 233 relevant cases by the above methods. These calculations include particles of up to 1 μm. In later figures and tables, we will use EAA-derived data where available during the diffusion sampling period.

It should be noted that the diffusion processor as employed in this study probably excluded a major portion of aerosol particles above about 5 μm from the unprocessed DBU aerosol samples; the significance of this will be discussed in a later section. There is also considerable uncertainty whether ideal diffusion battery behavior may be assumed for particles larger than 0.3 μm diameter.[18]

RESULTS AND DISCUSSION

Sulfate Data from High-Volume Sampling

Sulfate concentrations from the BNL 6 min/hr high-volume sampler, the Universal 24-hr high-volume sampler and the sum of sulfate concentrations on all five stages of the Andersen impactor for August, 1976 are reported in histographic form in FIGURE 2. We note the cyclic nature of the sulfate concentration provided as was observed for mass concentrations by other NYSAS collaborators.[19] There are clearly some significant differences between the observed data by the different sampling methods and other data. The BNL HiVol data and other data for 8/9–10 are explainable in terms of the occurrence of Hurricane Belle during this sampling period; the resultant marine aerosol apparently interfered with turbidimetric determination of sulfate but not the BaSO$_4$ gravimetric determination. Further comparison of data consisted of linear regression plots of BNL HiVol sulfate vs. Universal 24-hr HiVol and total cascade impactor data; the following regressions were obtained:

$$X = 0.645\,Y + 4.68, r = 0.847 \text{ (22 days)}$$
$$X = 0.755\,Z + 1.61, r = 0.802 \text{ (20 days)}$$

where X = BNL HiVol sulfate, Y = Universal HiVol sulfate and Z = total cascade impactor sulfate, all concentrations in μg/m^3. The correlation plot of X vs. Z suggested that data from 2 days' samples were inappropriately skewing the linear regression plot. Excluding those data resulted in the following regression line:

$$X = 0.938\,Z + 5.45, r = 0.883 \text{ (18 days)}$$

The slope indicates that, although the data are highly correlated, the Universal HiVol and total impactor concentrations are about 10% higher than the BNL HiVol data. It is certainly unlikely that the 6 min/hr sampling regime for the latter sampler would lead to these significant differences in sulfate concentrations over the same temporal scale. It is probable that artifact sulfate formation [20] on the Universal glass fiber filters and on the impactor plates and afterfilter led to elevated sulfate values for these samplers. It has been shown that the H_3PO_4-treated quartz filters used with the BNL HiVol sampler do not lead to "artifact" sulfate formation.[11] Although this explanation is in the form of an hypothesis, no convincing alternatives are evident to the authors.

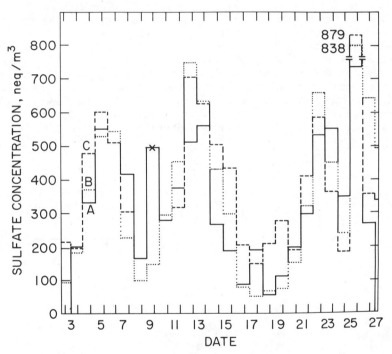

FIGURE 2. NYSAS HiVol sulfate data; A=BNL 6 min/hr HiVol; B=universal 24-hr HiVol; C=total Andersen impactor.

Chemical Composition Data from BNL HiVol Sampling

In addition to the soluble sulfate data reported above for BNL 6 min/hr HiVol samples, total sulfur analyses were performed on five of the samples by the [110]Ag method. Total sulfur values were lower by an average of about 15% than sulfate, although the r^2 correlation did exceed 0.9. Since the precision of the sampling and analysis by either method is about ±6 to 8%, this difference is marginally significant and in the direction reported by Forrest and Newman.[21] Attempts were made to determine strong acid content by

Gran titration for these HiVol samples. However, for all but three of the samples some or all of the leach solution was neutralized by dissolved particle constituents leading to a negative value for the strong acid concentration. As noted below, this was not observed for the diffusion processor samples. Hence, we conclude that basic, water-soluble constituents in the coarse particle regime generally may confound acid determinations in non-size-discriminated particle samples, especially those collected in urban areas.[11]

A much more fruitful comparison of soluble ion data in the BNL HiVol samples is suggested by FIGURE 3, which is a linear regression plot of particulate ammonium vs. turbidimetric sulfate concentrations, both expressed in units of equivalents ($\times 10^9$) per m^3. An exceptionally high correlation coefficient, $r = 0.926$, indicates that more than 85% of sulfate variability is predictable from the ammonium levels. This is especially intriguing since the slope of the regression plot indicates that about 45% of the sulfate appears to be associated with cations other than ammonium. The high correlation between ammonium and sulfate levels has been observed previously,[2, 11, 22] most recently in both urban and rural areas of southern Arizona.[23] Nevertheless, observation of this correlation of ammonium with sulfate in the aerosol of a large, industrial, urban area, in consideration of the presumed association of ammonium sulfate aerosol with formation and transport of sulfate on a regional scale, make this finding an important one.

Diffusion Processor Chemical Composition Data

The body of data from the diffusion processor for the period 8/3—8/26/76 is shown in histographic form in FIGURE 4. By comparison with the HiVol

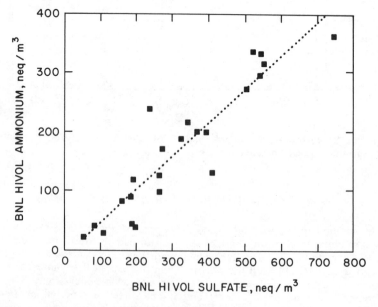

FIGURE 3. NYSAS correlation plot BNL HiVol ammonium vs. BNL HiVol sulfate; concentration units=neq/m^3.

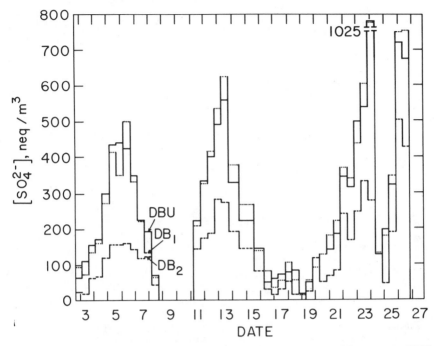

FIGURE 4. NYSAS diffusion sampling sulfate data, August, 1976; ———=DBU samples; . . . . .=DB1 samples; - - - - -=DB2 samples.

data in FIGURE 2, it can be seen that the trends in sulfate concentration are quite similar to the HiVol trends, although the finer time resolution in the diffusion battery data accentuates the temporal scale of changes in sulfate levels. Furthermore, it is clear that a major fraction of aerosol sulfate is removed by the diffusion processor from DB2 samples although not from DB1 samples relative to unprocessed aerosal samples, DBU. This will be discussed in more detail in a later section.

The chemical composition data from the associated cations, strong acid, and ammonium are plotted in equivalent units along with sulfate data for DBU samples in FIGURE 5. The trends in $[NH_4^+]$ and $([NH_4^+] + [H^+])$ were remarkably similar to that for sulfate.

Comparison regression plots of ammonium and sulfate in the diffusion-processed samples were made as for the HiVol data, and the linear regression plot for DB1 samples (particles below *ca.* 0.035 μm removed) is shown in FIGURE 6. Here again an exceptionally high correlation was observed ($r = 0.94$) with two principal differences—the x-intercept (representing sulfate not associated with ammonium) is an order of magnitude lower than for HiVol samples and, more significantly, the regression slope is 0.76 compared to 0.56 for HiVol data. The lower x-intercept indicates that a major portion of the coarse sulfate-containing particles are removed by the diffusion processor but are sampled by the HiVol sampler. This is confirmed by the fact that the average sulfate concentrations over 21 days on the HiVol samples was 16.6 μg/m³ compared to 13.4 μg/m³ for the unprocessed, diffusion-processor sam-

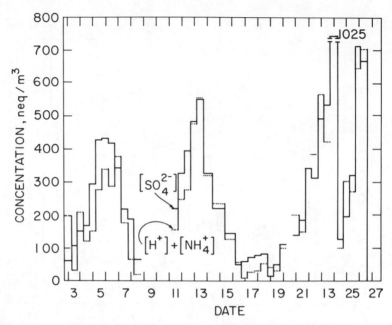

FIGURE 5. NYSAS sulfate chemical composition data from diffusion processor samples, August, 1976; ———— = sulfate concentration in DBU samples; . . . . . = sum of H^+ and NH_4^+ concentrations in DBU samples.

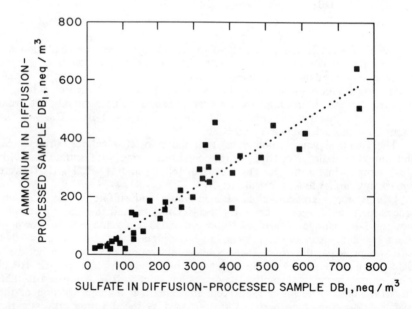

FIGURE 6. NYSAS correlation plot, ammonium vs. sulfate in DB1 diffusion processor samples, August, 1976.

ples and, in addition, by the fact that a plot of Δ(HiVol-DBU) ammonium concentration vs. Δ(HiVol-DBU) sulfate has a zero slope within experimental error-ergo, no coarse particle ammonium.

The higher slope of the NH_4^+ vs. SO_4^{2-} regression plot may also be related to the absence of coarse particle ammonium since, with a major portion of the coarse particle sulfate removed, a higher slope is to be expected. Essentially, the same slope and correlation coefficients for NH_4^+ vs. SO_4^{2-} regression plots were observed for the other diffusion-processed sample data-sets, DBU and DB2. In an attempt to determine if the NH_4^+/SO_4^{2-} ratio varied with particle size, a regression plot of the difference concentrations, $\Delta[NH_4^+]$ vs. $\Delta[SO_4^{2-}]$ between DBU and DB2 samples, was obtained which isolated the particle size region below *ca.* 0.3 μm. A correlation coefficient, r = 0.89 was obtained but the regression slope was only 0.66, marginally significantly lower than for each separately. If one assumes that the principal cationic component other than

TABLE 1

SUMMARY OF MULTIPLE REGRESSION ANALYSES

(NYSAS diffusion sampling data, New York, August, 1976)

$$Y = B_1X_1 + B_2X_2 + C$$

Y = dependent variable = $[SO_4^{2-}]$ by turbidimetry, neq/m³
X_1 = independent variable (1) = $[H+]$ by Gran titration, neq/m³
X_2 = independent variable (2) = $[NH_4^+]$ by indophenol, neq/m³

Sample Group	Regression Equation	Multiple r	$\sigma(B_1)$	$\sigma(B_2)$	$\sigma(C)$
DBU	$Y = 0.57\,X_1 + 1.03\,X_2 + 42$	0.950	0.17	0.07	15
DB1	$Y = 0.43\,X_1 + 1.11\,X_2 + 34$	0.945	0.37	0.08	17
DB2	$Y = 0.14\,X_1 + 1.08\,X_2 + 21$	0.924	0.27	0.08	11
(DBU-DB2)	$Y = 0.44\,X_1 + 0.86\,X_2 + 32$	0.913	0.16	0.07	10

ammonium associated with the diffusion-processed samples is H+, one would conclude that sulfate-containing suboptical particles (removed from sample DB2 by the diffusion processor) are probably somewhat more acidic than larger particles within the respirable size range.

To further elucidate this possibility, we obtained multiple regression analyses in which best fit regression lines were obtained for the dependent variable, sulfate, vs. the independent variables, strong acid (by Gran titration) and ammonium. The resultant data are listed in TABLE 1 for all three sets of diffusion-processor samples and for the suboptical fraction, DBU-DB2, by difference. It is clear that all of the regression slopes, B_2, for ammonium become insignificantly different from unity when the strong-acid concentrations are included in the regression plot. This strongly implies that ammonium is associated only with sulfate in these fine fraction, urban aerosol samples. In contrast, the regression slopes for strong acid in these samples do not become unity and there is much more scatter in the data as indicated by higher values for the standard error parameter, $\sigma(B_1)$. The regression slopes and correlation coefficients for strong-acid data from sample groups DBU, DB1, and (DBU-DB2) had values of approximately 0.5, which indicates that nearly all the sulfate variability can be accounted for by the ammonium and strong-acid variations, but that New York City summer aerosol contains other acid components not associated with sulfate and that the concentration(s) of these component(s) show more temporal variation than do ammonium sulfate levels.

A comparison of the daytime and nighttime ammonium and acid data indicated no significant diurnal variation in acid/sulfate and ammonium/sulfate ratios in either the optical or suboptical size range. In addition, no diurnal variation of ammonium or sulfate in HiVol samples could be observed.

Size Distribution Data for Sulfate

Andersen Cascade Impactor. A summary of the mean size distribution information obtained from thirty-eight 24 hr Andersen impactor samples collected at a flow rate of 42 cfm is listed in TABLE 2. For this data set the observed average sulfate concentration from all five stages was 19.5 $\mu g/m^3$. It can be seen that about 85% of the measured sulfate is present in what may be referred to as the fine particle range (<2.0 μm) with about 15% of the sulfate in the coarse particle range.

Assuming that the lower size limit for significant sulfate mass is 0.1 μm and realizing that the sampler employed probably exhibits severe sampling losses for particles >15 μm,[23] we have calculated an average mass distribution for sulfate from the above impactor data (FIGURE 7). Included with the average distribution is a distribution for an exceptionally clean day (8/17/76, $[SO_4^{2-}]_{total} = 7.3$ $\mu g/m^3$) and for a severely polluted day (8/25/76, $[SO_4^{2-}]_{total} = 42.2$ $\mu g/m^3$). No distinct coarse-particle mode is obvious from the average distribution, although individual distributions for both clean and polluted days do indicate such a mode. As noted by Knutson *et al.*,[3] the period August 25–29 was characterized by an exceptionally large volume in the 1–2 μm size region and this was reflected in the large $\Delta M/\Delta \log d_p$ value for sulfate in this size regime (see Curve C of FIGURE 7), ergo, 19% of the 42.2 $\mu g/m^3$ of sulfate was collected on Stage 4 of the impactor. The preponderance of sulfate in the submicrometer size range, in which range it constitutes at least 20% of the

TABLE 2

SUMMARY OF ANDERSEN IMPACTOR DATA

(Size distribution data for sulfate, NYSAS, New York, August, 1976)

Andersen Stage i	Size Range, Stage i, μm	Mean Fraction Sulfate Mass in Stage i	Standard Deviation of Mean	$-\sigma$ to $+\sigma$ Range, %
1	7–(15)	0.044	±0.046	0–9
2	3.3–7	0.077	±0.076	0–15
3	2.0–3.3	0.056	±0.059	0–11
4	1.1–2.0	0.104	±0.077	2–18
5 *	(0.1)–1.1	0.727	±0.155	57–88

* Back-up filter.

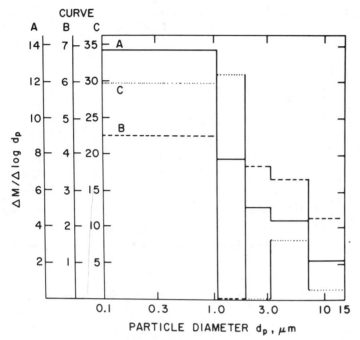

FIGURE 7. NYSAS sulfate size distributions from Andersen impactor data: A= average distribution for 20 days; B=distribution for 8/17/76; C=distribution for 8/25/76.

total mass, is, of course, not particularly surprising since many workers have observed this in aerosols from North American,[25] European,[26] and Japanese locations.[27]

Diffusion Sampler. It is most informative to compare the fraction of aerosol volume removed by diffusion-processed samples DB1 (d- ; = 0.035 μm) and DB2 ($d_{50\%}$ = 0.13 μm) with the corresponding aerosol volume or sulfate mass in the unprocessed aerosol samples, DBU. No significant sulfate mass or aerosol volume is found in Aitken nuclei removed prior to DB1 sample collection. However, as shown in FIGURE 8, a significant portion of the aerosol volume (about 20%) is removed from samples DB2 indicating that about 20% of the aerosol volume results from particles in the suboptical particle size regime below about 0.25 μm in diameter. In contrast, a range of 30–70% of the aerosol sulfate is similarly deduced to be in the suboptical particle size region. One may conclude that since a much larger portion of sulfate mass than total aerosol is in the suboptical region, the sulfate portion of the total aerosol increases with decreasing particle diameter within the aerosol volume mode at about 0.2 μm.

The average size distribution data observed for both DB1 and DB2 samples are summarized in TABLE 3. Clearly, sulfate constitutes a major portion of the submicron aerosol and its portion of the total aerosol increases with decreasing particle diameter to below the size cutoff for optical scattering techniques. The

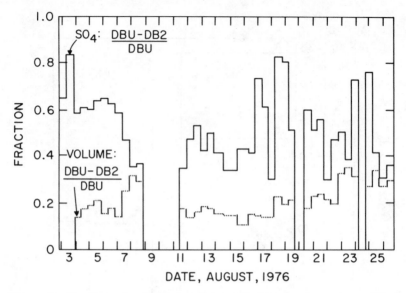

FIGURE 8. Fraction of suboptical sulfate and aerosol volume from diffusion-processed samples, NYSAS, August, 1976.

aerosol and sulfate size distribution data observed in New York during August, 1976, are in remarkably close agreement with a set of data obtained in rural Glasgow, IL (100 km of St. Louis, MO) during July, 1975.[4] In that rural environment about 20% of the aerosol volume and 50% of the sulfate mass was in the suboptical size region. This similarity of data on submicron aerosol from rural and urban locations is strong evidence for the contention[28] that a large portion of ambient aerosol sulfate is both formed and transported hundreds of kilometers downwind from the principal source regions of sulfate precursors.

TABLE 3

SUMMARY OF AEROSOL AND SULFATE REMOVAL IN DIFFUSION-PROCESSED SAMPLES

(NYSAS, New York, August, 1976)

Ratio	Aerosol Volume, $\frac{(\mu m)^3}{cc}$	Concentration of Sulfate, $\mu g/m^3$
DB1/DBU *	0.96±0.01	0.95±0.15
DB1/DBU *	0.78±0.04	0.95±0.15

* DBU: untreated aerosol; DB1: $d_{50\%} = 0.03$ μm; DB2: $d_{50\%} = 0.12$ μm.

CONCLUSIONS

Aerosol samples from a New York City site have been collected by a variety of sampling techniques during August, 1976 and analyzed for sulfate and related cationic species as a function of particle size from >7 μm to <0.05 μm diameter. The most significant results apparent from this extensive body of data are the following:

1. About 85% of the sulfate mass was found in respirable particles less than 2 μm in diameter; over 70% of the sulfate was found in submicron particles and at least 50% of the sulfate in particles below about 0.25 μm diameter.

2. The sulfate proportion of the aerosol volume increases with decreasing particle diameter throughout the respirable particle size range.

3. Ammonium in aerosol particles is highly correlated with sulfate concentration in respirable particle size ranges with correlation coefficient, $r = 0.92 - 0.94$. Multiple correlation of sulfate with strong acid and ammonium increases the value of r to >0.95, but there is an indication of other variable sources of strong acid which somewhat confound the data.

SUMMARY

Samples of airborne particles were collected in New York City during August, 1976 in connection with the New York Summer Aerosol Study (NYSAS) as follows: (1) 12-hr, low-volume samples were collected during daylight and darkness hours using a sampler based on the Sinclair diffusion battery; (2) 24-hr, 6 min/hr high-volume samples were collected on four in circles of acid-treated quartz filters; and (3) 24-hr high-volume samples were collected on conventional 8×10 in glass-fiber filters and Andersen impactor apparatus. Extracts of low volume samples (a) were analyzed and data for strong acid, ammonium and sulfate are herein reported. High correlation of ammonium and sulfate was observed in all fine particle size fractions. Fractions of sulfate in the nuclei and suboptical size ranges were remarkably close to values reported for rural ambient aerosols. Diurnal variations in sulfate and associated cations were insignificant. Speciated sulfate determinations from HiVol samples (b) were compared with data from low-volume samples (a) and with sulfate data from HiVol samples (c) obtained by two other laboratories using $BaSO_4$ gravimetric techniques. Agreement between these data was satisfactory considering the known artifact sulfate formation associated with collection on glass-fiber filters.

ACKNOWLEDGMENTS

We would like to thank our other collaborators in the New York summer aerosol study and, in particular, Morton Lippmann and Ted Kneip and their associates of New York University Medical Center for providing the principal sampling site and the personnel to operate it. The Brookhaven authors would like to thank D. Sinclair of U.S. D.O.E.'s Environmental Measurements Laboratory (formerly HASL) for the loan of the diffusion battery, our analytic staff (especially M. Phillips) for performing the analyses, and J. Tichler and

C. Saurino for assistance in utilizing the SPSS programs for regression analyses. One of us (B. L.) acknowledges partial support for this work provided by the Department of Epidemiology and Public Health, Yale University School of Medicine through Grant No. 5-T01-ES-00123–08, the John B. Pierce Foundation Laboratory through Grant No. 5-S07-RR-05692–07, and the New York State Department of Environmental Conservation.

REFERENCES

1. KNEIP, T. J., B. P. LEADERER, D. M. BERNSTEIN & G. T. WOLFF. 1979. The York Summer Aerosol Study (NYSAS), 1976. Ann. N.Y. Acad. Sci. **322**. This volume.
2. MARLOW, W. H. & R. L. TANNER. 1976. Diffusion sampling method for ambient aerosol size discrimination with chemical composition determination. Anal. Chem. **48**: 1999–2001.
3. KNUTSON, E. O., D. SINCLAIR & B. P. LEADERER. 1979. New York summer aerosol-number concentration and size distribution of atmospheric particles. Ann. N.Y. Acad. Sci. **322**. This volume.
4. TANNER, R. L. & W. H. MARLOW. 1977. Size discrimination and chemical composition of ambient airborne sulfate particles by diffusion sampling. Atmos. Environ. **11**: 1143–1150.
5. ANDERSEN, A. 1966. A sampler for respiratory health hazard assessment. Am. Ind. Hyg. Assoc. J. **27**: 160–165.
6. MULLIN, J. B. & J. P. RILEY. 1955. The spectrophotometric determination of nitrate in natural waters, with particular reference to sea-water. Anal. Chim. Acta **12**: 464–480.
7. KLEINMAN, M. T., B. P. LEADERER, C. TOMCZYK & R. L. TANNER. 1979. Inorganic nitrogen compounds in New York City air. Ann. N.Y. Acad. Sci. **322**. This volume.
8. HANSEN, L. D., L. WHITING, D. J. EATOUGH, T. E. JENSEN & R. M. IZATT. 1976. Determination of sulfur (IV) and sulfate in aerosols by thermometric methods. Anal. Chem. **48**: 634–638.
9. HANSEN, L. D., R. M. IZATT, M. W. HIL, N. F. MANGELSON, J. J. CHRISTENSEN & D. J. EATOUGH. 1977. Reduced species and acid-base components of the New York City aerosols. Presented at the American Industrial Hygiene Conference, New Orleans, LA, May 22–27.
10. ASKNE, C. & C. BROSSET. 1972. Determination of strong acid in precipitation, lake-water and airborne matter. Atmos. Environ. **6**: 695.
11. TANNER, R. L., R. CEDERWALL, R. GARBER, D. LEAHY, W. MARLOW, R. MEYERS, M. PHILLIPS & L. NEWMAN. 1977. Separation and analysis of aerosol sulfate species at ambient concentrations. Atmos. Environ. **11**: 955–966.
12. KEAY, J. & P. M. A. MENAGE. 1969. Automated distillation procedure for the determination of nitrogen. Analyst (London) **94**: 895–899; ibid. 1970. **95**: 379–382.
13. Sulfate method VIb by turbidimetry. 1959. Technicon Corporation, Tarrytown, N.Y.
14. FORREST, J. & L. NEWMAN. 1973. Sampling and analysis of atmospheric sulfur compounds for isotope ratio studies. Atmos. Environ. **7**: 561–573.
15. HUSAR, J. D., R. B. HUSAR & P. K. STUBITS. 1975. Determination of submicrogram amounts of atmospheric particulate sulfur. Anal. Chem. **47**: 2062–2065.
16. TANNER, R. L., R. GARBER & L. NEWMAN. 1977. Speciation of sulfate in ambient aerosols by solvent extraction with flame photometric detection. Paper ENVR-41, 173rd National Meeting, American Chemical Society. New Orleans, I.A., March 20–25.
17. TWOMEY, S. 1975. Comparison of constrained linear inversion and an iterative

nonlinear algorithm applied to the indirect estimation of particle size distributions. J. Comp. Phys. **18:** 188–200.

18. MARLOW, W. H. 1977. Optical-size particle penetration through a diffusion processor for filter sampling. Presented at the American Industrial Hygiene Conference, New Orleans, LA., May 22–27.

19. LIPPMANN, M., M. T. KLEINMAN, D. M. BERNSTEIN & B. P. LEADERER. 1979. Size-mass distribution of the New York summer aerosol. Ann. N.Y. Acad. Sci. **322**. This volume.

20. PIERSON, W. R., R. H. HAMMERLE & W. Y. BRACHACZEK. 1976. Sulfate formed by interaction of sulfur dioxide with filters and aerosol deposits. Anal. Chem. **48:** 1579–1584.

21. FORREST, J. & L. NEWMAN. 1977. Silver-110 microgram sulfate analysis for the short time resolution of ambient levels of sulfur aerosol. Anal. Chem. **49:** 1808–1811.

22. STEVENS, R. K., T. G. DZUBAY, G. RUSSWORM & D. RICKEL. 1978. Sampling and analysis of atmospheric sulfate and related species. Atmos. Environ. **12:** 55–68.

23. MOYERS, J. L., L. E. RANWEILER, S. B. HOPF & N. E. KORTE. 1977. Evaluation of particulate trace species in southwest desert atmosphere. Environ. Sci. Tech. **11:** 789–795.

24. WEDDING, J. B., A. R. McFARLAND & J. E. CERMAK. 1977. Large particle collection characteristics of ambient aerosol samplers. Environ. Sci. Technol. **11:** 387–390.

25. DZUBAY, T. G. & R. K. STEVENS. 1975. Ambient air analysis with dichotomous sampler and X-ray fluorescence spectrometer. Environ. Sci. Technol. **9:** 663–669.

26. BROSSET, C., K. ANDREASSON & M. FERM. 1975. The nature and possible origin of acid particles observed at the Swedish west coast. Atmos. Environ. **9:** 631–642.

27. KADOWAKI, S. 1976. Size distribution of atmospheric total aerosols, sulfate, ammonium and nitrate particulates in the Nagoya area. Atmos. Environ. **10:** 39–43.

28. WHITE, W. H., J. A. ANDERSON, D. L. BLUMENTHAL, R. B. HUSAR, N. V. GILLANI, J. D. HUSAR & W. E. WILSON, JR. 1976. Formation and transport of secondary air pollutants: ozone and aerosols in the St. Louis urban plume. Science **194:** 187–189.

INORGANIC NITROGEN COMPOUNDS IN
NEW YORK CITY AIR

Michael T. Kleinman * and Carol Tomczyk

*Institute of Environmental Medicine
New York University Medical Center
New York, New York 10016*

Brian P. Leaderer

*Department of Epidemiology and Public Health
John B. Pierce Foundation Laboratories
Yale University School of Medicine
New Haven, Connecticut 06519*

Roger L. Tanner

*Atmospheric Sciences Division
Department of Energy and Environment
Brookhaven National Laboratory
Upton, New York 11973*

INTRODUCTION

As part of a multilaboratory cooperative experiment, several facets of the complex interrelationships between nitrogen dioxide (NO_2), ammonium ion (NH_4^+), and nitrate ion (NO_3^-) were explored on air samples collected during August 1976 in New York City.

There has been considerable recent interest in understanding the relationships between source emissions of nitrogen compounds, their ambient atmospheric concentrations, and their impact on the environment and human health. This is especially true of NO_2, since significant health benefits might be expected from maintaining levels below the Federal Air Quality Criterion of 0.05 ppm.[1]

Most of the previous literature describing the sources and particle size distributions of inorganic nitrogen compounds has been based on measurements in the air of other cities. It is well known, for example, that nitrogen oxides (NO_x), principally nitric oxide (NO) and nitrogen dioxide (NO_2), are major components of effluents from combustion processes, i.e., space heating, power generation, internal combustion engines, and incineration.[2,3] Most of the ammonia in ambient atmospheres comes from decay of biologic materials; however, as much as 10% may be accounted for by automobiles.[4] Through a variety of chemical and photochemical reactions, some of which involve other

* Current affiliation:
    Environmental Health Laboratory
    Rancho Los Amigos Hospital
    University of Southern California
    Downey, California 90242.

115

0077-8923/79/0322-0115 $01.75/0 © 1979, NYAS

pollutants such as ozone (O_3) and reactive hydrocarbons, gaseous NO_x can be converted to particulate nitrates or nitrites. One might therefore expect to see a strong correlation between NO_2 and nitrates.

In many cities, nitrates have been reported as occurring on small particles; [5, 6] however, Novakov et al.,[7] reported that in Pasadena a large fraction of nitrates were found on large particles. Miller et al.[8] and Kadowaki[9] have reported findings similar to those of Novakov for New York and Nagoya, Japan, respectively. Harker et al.[10] have demonstrated in the laboratory that photochemical oxidation of SO_2 to $SO_4^=$ decreased the amount of nitrate in photochemically produced aerosol. In the above study, sulfuric acid reacted with aerosol ammonium nitrate forming a bisulfate salt, thus preventing loss of NH_4^+ from the particles, while releasing nitric acid vapor from the aerosol sample.

Harker[10] predicted on the basis of above results that eastern urban centers with high ambient SO_2 levels, on the order of 0.05 ppm, should exhibit significant suppression of the nitrate content of their photochemical aerosols.

In general, the results obtained during the 1976 New York Summer Aerosol Study (NYSAS) bear out Harker's prediction. As will be shown in this report, nitrates in New York City aerosol were found predominantly associated with particles of mass median aerodynamic diameter (MMAD) of about 5 μm, and were generally uncorrelated with gaseous precursor compounds.

METHODS AND MATERIALS

Total suspended particulate matter (TSP) (24-hour Hi-Vol sample) was collected using a standard high volume air sampler. Twenty-four hour size-fractionated samples were obtained using an Andersen 2000 high volume cascade impactor. A diffusion battery sampler, which provided samples of ultrafine aerosols, was run for 12-hour intervals, corresponding to daylight and night-time hours. This sampler, which has 50% "cutoffs" at about 0.035 and 0.15 μm diameter, has been described by Marlow et al.[11] A 50% cutoff in this case means that 50% of the particles of the stated size are removed prior to the sampling port, and 50% penetrate to the filter. Twenty-four hour integrated NO_2 samples were collected using a personal diffusion sampler.[12] A breakdown of the sampling schedule, laboratories, analytic methods and type of data provided is shown in TABLE 1.

Traffic count data were provided by the New York City Traffic Department. Sensors were placed on the East River Drive, between 32nd and 31st Streets. Because of sporadic breakdown in the traffic-counting equipment, it was often necessary to interpolate missing data points. A linear interpolation between available hourly traffic averages was used, and we feel that the resulting 24-hour traffic totals are fairly representative of the actual highway values, even though complete hourly data were not available for about half of the days.

Concentrations of trace metals in 24 hour hi-vol samples were also determined during NYSAS-1976, as reported by Bernstein et al.[13] These data, along with meteorologic information provided by the U.S. National Weather Service, were used to investigate the relationships between ambient concentrations of nitrogen species and emissions from pollution source classes in New York City.

RESULTS

The concentrations of NO_2 in ppm, NH_4^+, and NO_3^- in $\mu g/m^3$, the fraction of the total nitrate associated with particles less than 3.5 μm MMAD, expressed in percent, and 24-hour traffic counts observed during NYSAS-1976, are shown in TABLE 2. The concentrations of NO_2, NH_4^+, and NO_3^-, ranged from 0.04 to 0.11 ppm, 0.14 to 6.50 $\mu g/m^3$, and 0.46 to 4.19 $\mu g/m^3$ respectively. Traffic, on the other hand, varied over a relatively small range, 76 to 139 thousand cars per day. From these data alone, one could infer that changes in traffic could

TABLE 1

INORGANIC NITROGEN COMPOUNDS

Data Sources	METHODS
A. Brookhaven National Laboratory	
1. Diffusion Battery Processed	
NO_3^- Size-fractionated samples	Auto analyzer (hydrazine reduction, colorimetric)
NH_4^+ 12-hr samples	Indophenol, colorimetric
2. Hi-Vol	
NO_3^- Total aerosol	Auto analyzer (hydrazine reduction, colorimetric)
NH_4^+ 24-hr	Indophenol, colorimetric
B. Yale University	
1. Hi-Vol	
NO_3^- 24-hr	Hydrazine reduction, colorimetric
2. Andersen Hi-Vol	
NO_3^- 24-hr	Hydrazine reduction, colorimetric
C. New York University	
1. Hi-Vol	
NO_3^- 24-hr	Cadmium column reduction, colorimetric
NO_2^- 24-hr	No reduction—colorimetric
Trace metals	Acid dissolution—atomic absorption
2. NO_2 Gas Sampler 24-hr	Adsorption in triethenolamine—colorimetric

account for only a small part of the concentration variations observed for nitrogen species.

Temporal patterns observed for NO_2, NH_4^+ and total NO_3^- are shown in FIGURE 1. The data presented were normalized with respect to the dispersion factor (the product of morning mixing depth and wind speed averaged through the mixing depth, divided by a long-term average product of 4200 $m^2 sec^{-1}$).[14, 15] Normalizing allows one to adjust for day-to-day changes in the dispersive properties of the atmosphere, so that underlying changes in pollutant concentrations can be more easily distinguished.

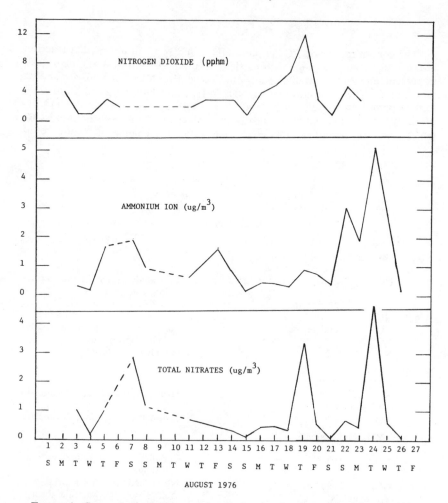

FIGURE 1. Concentrations of inorganic nitrogen species (normalized for dispersion).

Although there seem to be some relationships among these data, for example, peak values for the variables occurring on the 19th, the 22nd and the
24th, the peak heights do not change proportionally. As is common in environmental field studies, there are "breaks" in the data due to sample losses or
sampler malfunctions, and these complicate efforts to determine the meaning-
fulness of observed relationships. For this month of study, there were two to
five days for which data were not obtained, depending on the variable examined.

The particle size data obtained using the Andersen 2000 impactor, indicate
that, on the average, 58% of the nitrates are associated with particles greater
than 3.5 μm MMAD. It is possible to obtain three additional points in a size
distribution for nitrate aerosols from the diffusion battery and hi-volume air

sampler data. This distribution is shown in FIGURE 2. These data indicate that the MMAD for nitrate is about 5 μm, which is in good agreement with that observed with the Andersen 2000 sampler, about 4.4 μm. This confirmation is important because it suggests that the apparent large NO_3^- particle size is not an artifact due to some characteristic of the impactor samples.

The relationships between possible source emissions and atmospheric concentrations of inorganic nitrogen species is obviously complex; however, it is often possible to uncover strong links by examining correlation coefficients, arranging these in a matrix, and by looking for underlying patterns among those variables exhibiting significant correlation. Such a matrix (TABLE 3) was constructed for NO_2, total NO_3^-, and fine particle NO_3^- along with traffic counts, and concentrations of Pb, which is a major constituent of automotive emissions, and V, which is a trace constituent of particles emitted from oil burners.

The coefficients indicate that traffic, Pb and V correlate strongly with NO_2 and that weaker relationships exist between V, NO_2 and fine particle NO_3^-. A possible conceptualization of the observed relationships is shown in FIGURE 3. Traffic emissions produce Pb as well as NO_2, but do not contain significant concentrations of nitrates. Fuel oil burning, on the other hand, produces fine particles enriched in V which also either contain NO_3^- indigenously or which in some way become associated with NO_3^-, for example by conversion of NO_2 to NO_3^- on the surface of the particles. This conceptual framework is, of course, hypothetical. It is certainly not the only possible explanation of the month's observations, and further studies will be necessary to substantiate the above hypothesis.

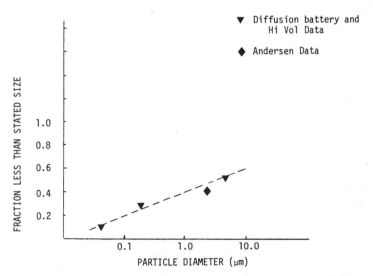

FIGURE 2. Nitrate particle size distribution data. Diameters are for 50% cuts and are not absolute size values.

DISCUSSION

New York City has kept recent atmospheric SO_2 levels very low by strict enforcement of regulations prohibiting the burning of high sulfur fuels. Average

TABLE 2

CONCENTRATIONS OF INORGANIC NITROGEN COMPOUNDS
IN NEW YORK CITY AIR MEASURED DURING AUGUST 1976

Date	24-Hour Traffic Counts (thousands of cars)	24 Hour Average			Fine Particle $NO_3^=$ ‡ (% ≤3.5 μm)
		NO_2* (ppm)	NH_4^+ † (μg/m³)	$NO_3^=$ † (μg/m³)	
8/2–3/76	—	—	—	—	—
8/3–4/76	—	0.089	0.75	2.15	50
8/4–5/76	—	0.089	3.42	3.66	50
8/5–6/76	—	0.044	5.30	3.29	50
8/6–7/76	139	0.021	—	—	—
8/7–8/76	107	0.027	2.40	3.59	40
8/8–9/76	85	—	1.49	1.74	20
8/9–10/76	76	—	0.14	1.56	60
8/10–11/76	81	0.076	3.12	1.56	—
8/11–12/76	128	0.106	3.64	4.19	—
8/12–13/76	116	0.119	4.91	2.63	30
8/13–14/76	112	0.084	5.65	1.66	20
8/14–15/76	102	0.067	2.30	0.87	20
8/15–16/76	93	0.067	1.63	0.95	40
8/16–17/76	110	0.050	0.76	0.69	50
8/17–18/76	112	0.076	0.82	0.84	60
8/18–19/76	105	0.087	0.44	0.46	40
8/19–20/76	114	0.073	0.56	2.02	40
8/20–21/76	112	0.096	2.16	1.46	50
8/21–22/76	—	0.107	4.28	0.53	50
8/22–23/76	—	0.110	6.02	1.40	40
8/23–24/76	130	0.105	5.96	1.19	40
8/24–25/76	111	—	3.91	3.50	40
8/25–26/76	121	—	6.50	1.81	40
8/26–27/76	108	—	1.79	0.61	40
8/27–28/76	112	—	—	—	40
8/28–29/76	91	—	—	—	50
8/29–30/76	77	—	—	—	50
8/30–31/76	101	—	—	—	50

* Gas diffusion sampler.
† BNL Hi-Vol.
‡ Andersen 2000 Impactor.

concentrations of SO_2 are generally below the federally mandated standard of ~80 μg/m³. During the summer of 1976, the mass of particulate nitrate was found to be predominantly associated with large particles. This association may be due to photochemical conversion of SO_2 to $SO_4^=$ particulates, with resulting removal of NO_3^- from fine particles, but such conversions probably

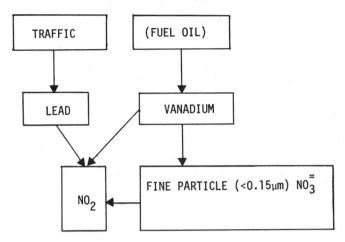

FIGURE 3. Observed relationships between sources and nitrogen oxides in New York City air.

did not occur within the New York City urban airshed.[16] It is also possible that photochemical reactions caused removal of nitrates from small particles resulting in an apparent shift of nitrate to large particles in New York City air, and those those reactions occurred in the air mass before the air mass reached New York City.

The problem of sampling artifacts should also be considered in interpreting our results. Several commonly used filter media have been shown to adsorb gaseous nitrogen compounds and yield erroneously high results. Spicer[17] has demonstrated that quartz fiber filters such as were used in the Brookhaven sampler, minimize such nitrate artifact information. Another difficulty is encountered in total particulate collections, because small ambient particles,

TABLE 3

OBSERVED INTERRELATIONSHIPS AMONG VARIABLES THOUGHT TO BE ASSOCIATED WITH AIRBORNE INORGANIC NITROGEN SPECIES

| | Correlation Coefficient, R (number of sample pairs) | | | | NO_3^- | |
	Traffic	Pb	V	NO_2	Total	Fine
Traffic	1.00					
Pb	$0.78_{(15)}$***	1.00				
V	$0.36_{(13)}$	$0.49_{(9)}$**	1.00			
NO_2	$0.60_{(12)}$**	$0.81_{(9)}$***	$0.91_{(9)}$***	1.00		
Total $NO_3^=$	$0.53_{(10)}$	$0.21_{(9)}$	$0.18_{(9)}$	$0.26_{(12)}$	1.00	
Fine Particle $NO_3^=$	$0.44_{(10)}$	$0.43_{(11)}$	$0.58_{(11)}$*	$0.62_{(11)}$**	$0.48_{(10)}$	1.00

* Probability that $R = 0 \leq 0.10$.
** Probability that $R = 0 \leq 0.05$.
*** Probability that $R = 0 \leq 0.01$.

which are usually acidic, might neutralize or acidify large, usually basic, particles and release HNO_3, which has a relatively high vapor pressure. This could result in erroneously low nitrate results.

With respect to our study, the above problems have been minimized by using quartz filters on the diffusion battery sampler (which collects size fractionated aerosols by filtration) and by separating, to some extent the large and small particles, which should minimize chemical interactions. Glass fiber filters were used in the Andersen impactor. Since one would expect little or no collection efficiency for vapor species by impaction, only the final filter stage might have artifactually high values. This would, however, have resulted in an apparently low particle size for nitrates rather than the large size found in this study. Artifact formation on impactor final filters might explain why low particle sizes were observed for nitrates in some previous studies.

CONCLUSIONS

Several laboratories participated in this cooperative study of inorganic nitrogen species using a variety of sampling and analytic techniques. This integrated approach allowed a relatively short-term and limited experiment to produce a fairly comprehensive conceptual picture of the behavior of inorganic nitrogen compounds in New York City air during the summer season. The data show that more than 50% of the aerosol nitrate is contained in particles greater than 5 μm diameter and that concentration variations of these large particles are not directly related to emissions by oil burners or autos, although the latter sources are strongly related to NO_2 concentrations. There is an indication that the concentration of fine particle nitrates is related to oil burning sources.

A possible explanation for the apparent low concentrations of nitrate in fine particles in New York City air is that photochemical conversion of SO_2 to $SO_4^=$ can result in a decrease of NO_3^- from photochemical aerosols during transport of an air mass over areas with relatively high SO_2 concentrations. Since photochemical processes generally only involve fine particles, the NO_3^- contents of the fine particles can be greatly reduced, while significant concentrations of NO_3^- remain in large particles.

ACKNOWLEDGMENTS

The authors wish to thank Marie Ann Leyko and Patricia Imbrosciano, who performed the nitrate analyses for NYU and Robert Mallon who handled the day-to-day sampling operations for this study. We also wish to thank Mr. Stanley Bogoff and his staff at the New York City Department of Traffic for providing the traffic data that were used in our data analyses.

The work of New York University is supported by Grant No. RP439–1 of the Electric Power Research Institute and the American Petroleum Institute and is part of a Center program supported by Grant No. ES 00260, from the National Institute of Environmental Health Sciences, and Grant No. CA 13343, from the National Cancer Institute.

Partial support for this work was also provided by the Department of Epidemiology and Public Health, Yale University, through Grant No. 5–101–ES–00123–08, the John B. Pierce Foundation Laboratory through Grant No.

5–SO7–RR–05692–7 and the New York State Department of Environmental Conservation.
The work at Brookhaven National Laboratory is supported by the Division of Biomedical and Environmental Research of the U. S. Department of Energy.

REFERENCES

1. LEADERER, B. P., R. T. ZAGRANISKI & J. A. J. STOLWIJK. 1976. Estimates of health benefits due to reductions in ambient NO_2 levels. Environ. Manag. **1:** 31.
2. SANDBERG, J. S., D. A. LEVAGGI, R. E. DEMANDEL & W. SUI. 1976. Sulfate and nitrate particulates as related to SO_2 and NO_x gases and emissions. J. Air Pollut. Control Assoc. **26:** 559.
3. URONE, P. 1976. The primary air pollutants—gaseous. *In* Air Pollution, Vol. I, 3rd edition. A. C. Stern, Ed. Academic Press. New York.
4. HARKINS, J. H. & S. W. NICKSIE. 1967. Ammonia in auto exhaust. Environ. Sci. Technol. **1:** 751.
5. LEE, R. E., JR. & R. K. PATTERSON. 1969. Size determination of atmospheric phosphate, nitrate, chloride and ammonium particulate in several urban areas. Atmos. Environ. **3:** 249.
6. HIDY, G. M. *et al.* 1975. Summary of the California aerosol characterization experiment. J. Air Pollut. Control Assoc. **25:** 1106.
7. NAVAKOV, T., P. K. MUELLER, A. E. ALCOCER & J. W. OTVOS. 1972. Chemical composition of Pasadena aerosol by particle size and time of day. J. Colloid Interface Sci. **39:** 285–294.
8. MILLER, D. F., W. E. SCHWARTZ, P. E. JONES, D. W. JOSEPH, C. W. SPICER, C. J. PIGGLE & A. LEVY. 1973. Haze formation: Its nature and origin. Battelle-Columbus Laboratories Report to EPA and CRC. EPA Report #650/3/74/ 002 NERC. Research Triangle Park, N.C., June 1973.
9. KADOWAKI, S. 1977. Size distribution and chemical composition of atmospheric particulate nitrate in the Nagoya area. Atmos. Environ. **11:** 671.
10. HARKER, A. B., L. W. RICHARDS & W. E. CLARK. 1977. The effect of atmospheric SO_2 photochemistry upon observed nitrate concentrations in aerosols. Atmos. Environ. **11:** 87.
11. MARLOW, W. H. & R. L. TANNER. 1976. Diffusion sampling method for ambient aerosol size discrimination with chemical composition determination. Anal. Chem. **48:** 1999.
12. PALMES, E. D., A. F. GUNNISON, J. DIMATTIO & C. TOMCZYK. 1976. Personal sampler for nitrogen dioxide. Am. Ind. Hyg. Assoc. J. **37:** 470.
13. BERNSTEIN, D. M., K. A. RAHN & B. LEADERER. 1979. New York summer aerosol study: Trace element concentrations vs. particle size. Ann. N.Y. Acad. Sci. **322.** (This volume.)
14. KLEINMAN, M. T., T. J. KNEIP & M. EISENBUD. 1976. Seasonal patterns of airborne particulate concentrations in New York City. Atmos. Environ. **10:** 9.
15. KLEINMAN, M. T., D. M. BERNSTEIN & T. J. KNEIP. 1977. An apparent effect of the oil embargo on total suspended particulate matter in New York City air. J. Air Pollut. Control Assoc. **27:** 65.
16. WOLFF, G. T., P. J. LIOY, B. LEADERER, D. M. BERNSTEIN & M. T. KLEINMAN: Characterization of aerosols upwind of New York City, I. Transport. Ann. N.Y. Acad. Sci. **322.** (This volume.)
17. SPICER, C. Y. 1977. Photochemical atmospheric pollutants derived from nitrogen oxides. Atmos. Environ. **11:** 1089.

THE NATURE OF THE ORGANIC FRACTION OF
THE NEW YORK CITY SUMMER AEROSOL *

Joan M. Daisey and Marie Ann Leyko

Institute of Environmental Medicine
New York University Medical Center
New York, New York 10016

Michael T. Kleinman

Environmental Health Laboratory
Rancho Los Amigos Hospital
University of Southern California
Downey, California 90242

Eva Hoffman

Graduate School of Oceanography
University of Rhode Island
Kingston, Rhode Island 02881

INTRODUCTION

Organic compounds constitute a significant fraction of suspended particulates in the urban aerosol both in terms of mass and in terms of possible environmental health effects. Many of the organic compounds which have been detected in urban aerosols have been shown to be carcinogens and co-carcinogens.[1,2] Hueper and co-workers[3] tested aliphatic, aromatic and oxidized hydrocarbon fractions of suspended particulates from eight cities for carcinogenic activity in mice and found that two aromatic and two oxidized hydrocarbon fractions induced tumors. Similar results were reported for organic solvent extracts of engine exhaust by Kotin and co-workers.[4] Wynder and Hoffmann[5] found tumor-promoting activity from a neutral oxidized hydrocarbon fraction of suspended particulates. Many of compounds suspected of being present, such as epoxides and peroxides, have been shown to be carcinogens and co-carcinogens in animals.[6]

Although there have been a number of studies directed toward the characterization of the organic fraction of the urban aerosol, relatively little work has been done in New York City. Those studies which have been done[7-11] have been generally limited in their measurements of materials in the aerosol other than selected organic compounds. Usually no more than two laboratories examined one or only a few parameters. The New York Summer Aerosol Study (NYSAS-1976) presented a unique opportunity to investigate the organic fraction of the aerosol and to relate the data to a large number of other variables which were simultaneously measured by other laboratories.

---

* Presented in part at the American Industrial Hygiene Association 1977 Conference, New Orleans, Louisiana, May 22–27, 1977, as part of The New York Summer Aerosol Study.

125

Total masses of nonpolar and polar organic compounds from 24-hour total suspended particulate matter (TSP) samples were determined by sequential solvent extraction with cyclohexane, dichloromethane, and acetone. Nonextractable carbon was determined on selected samples. Daily variations in the concentrations of organic solvent extractable materials and selected polycyclic aromatic hydrocarbon (PAH) compounds were determined and correlations with trace metals, sulfate, nitrate, NO_2, ozone, traffic and meteorological variables were investigated. Selected solvent extracts were tested for bacterial mutagenic activity using the Ames test.[12]

EXPERIMENTAL PROCEDURES

Sampling

Twenty-four hour high-volume samples of total suspended particulate matter were collected from noon of one day to noon of the next during two alternate weeks in August, 1976. The samples were collected on 8″ × 10″ pre-extracted (1-propanol) Gelman Type A fiberglass filters at flow rates of approximately 1.55 m^3/min. The average air volume was 2200 m^3. Upon removal from the filter-head, the loaded filters were stored in glass jars under argon at −20° C. Maximum storage time prior to extraction was 4 days. Total particulate matter on the filters ranged from 90 to 490 mg.

Extraction

Several workers [13, 14] have shown that a significant fraction of organic material is not removed when a single non-polar solvent such as benzene is used for the extraction of suspended particulate matter. In view of this, a sequential solvent extraction was employed with increasingly polar solvents, cyclohexane, dichloromethane, and acetone, used in that order. Eight-hour extractions using 150 ml of each solvent were carried out. Samples were processed in groups of three plus a blank. Extracts and filters were stored in a refrigerator, in the dark, overnight between extractions. Extracts were then filtered and reduced to 10 ml in volume using a rotary evaporator with a water bath maintained at 35–45° C. The round bottom flask used in this operation was rinsed with solvent to thoroughly remove all of the sample extract. This brought the evaporated sample to a 20 ml volume. Samples were then stored in a freezer until analyzed. As the PAH compounds are sensitive to ultraviolet light, all extractions, evaporations, etc. were carried out with the laboratory lights off, the light being that admitted through north windows.

The dichloromethane and acetone extracts were evaporated to dryness in small tared aluminum foil pans on a slide warmer at 40° C and then weighed on an analytic balance. The samples were redissolved in 5 ml of solvent and stored again in the freezer.

As quantitative analyses for PAH were to be carried out, the cyclohexane extracts were not evaporated to dryness to prevent the loss of more volatile compounds. Residue weights were determined by weighing the material in 50 μl aliquots on small foil pans taken to dryness at 40° C. These weights were determined on a Cahn Electrobalance. All extract weights were blank corrected.

Nonextractable Carbon

Three samples and a blank filter which had been sequentially extracted were analyzed for nonextractable carbon using the wet persulfate oxidation technique and a Total Organic Carbon Analyzer. The dried filters were heated in an oven at 100° C for 4 hours prior to carbon analysis to drive off any remaining traces of organic solvent. Four 1.5–2.0 cm² sections for each sample were analyzed and the results were averaged. Values ranged from approximately 15–50 µg of carbon per cm². The average for the blank filter was 0.4 µg/cm².

Polycyclic Aromatic Hydrocarbons

Polycyclic aromatic hydrocarbons (PAH) were analyzed using a thin-layer chromatography/gas chromatography (TLC/GC) method.[15] The cyclohexane extracts were separated into three PAH fractions by thin-layer chromatography and then analyzed on a dual column Varian Model 2740 Gas Chromatograph (GC) equipped with a 12′ × ⅛″ stainless steel column packed with 6% Dexsil 300 on 80/100 mesh Chromosorb W(HP) and two flame ionization detectors. Operating conditions were: Injector—325° C; Detector—325° C; Carrier gas—nitrogen, 40 ml/min; Temperature Program—165° C for 2 min; 165° C to 285° at 4° C/min; hold at 285° C for 10–30 minutes. *N*-eicosane or *p*-terphenyl was used as an internal standard. Immediately prior to GC analysis, a reduction of sample volume from about 20 ml to about 30–50 µl was made. One to three microliters of the sample was then injected for analysis.

Trace Metals

High volume samples of total suspended particulates which had been collected on 8″ × 10″ Whatman filters from noon to noon were used for trace metal analysis. Half of each filter was used in the analysis. The samples were analyzed for Cd, Fe, Mn, Pb, V, and Zn using atomic absorption spectrophotometry as described by Eisenbud and Kneip.[16]

Other Data

The investigation of correlation between the organic solvent extracts and other variables involved the use of data which were generously provided by several other investigators in NYSAS-1976. Ozone data were provided by the New York State Department of Environmental Conservation. Meteorological data from the National Weather Service (N.Y.C.) were also used. S. Bugoff of the New York City Department of Transportation provided traffic count data.

Statistical Analysis

Relationships between the ambient aerosol concentrations of the organic solvent extracts and trace metals, ozone, sulfate, nitrate and ammonium ion, nitrogen dioxide and minutes of sunlight per day were examined using a nonparametric statistical test, the Kendall correlation.[17] Correlation coefficients and the levels of significance were determined for each paired set of data. Hypotheses regarding the sources of organic pollutants were tested using multiple regression analysis.[18]

Solvent extract data were also grouped by wind direction for the date of collection and the average concentration of the materials from each of the solvent extracts for a given wind direction was calculated. Student's t-test was then used to determine whether the averages were significantly different.

Bacterial Mutagenicity Tests

Selected organic solvent extracts were assayed for mutagenic activity using the Ames *Salmonella typhimurium* test system, but without microsomal activation.[12] Thus, only direct acting mutagens are detected. The cyclohexane and dichloromethane extracts used in the bacterial mutagenicity testing were evaporated to dryness, in a hood, at 35–40° C and redissolved in acetone for testing.

RESULTS AND DISCUSSION

Organic Solvent Extracts

One of the primary objectives of this investigation into the nature of the organic fraction of the New York City aerosol was to obtain information on the total amounts of organic material present in suspended particulates in New York City. Concentrations of organic materials in aerosols are usually determined by extracting suspended particulate samples with organic solvent, evaporating the solvent after extraction, and weighing the residue. Benzene has been the most commonly used solvent in the past, although cyclohexane has also been used because its extraction efficiency is close to that of benzene [14] and it is less toxic. Grosjean [14] has recently shown that extractions with either of these nonpolar solvents alone may seriously underestimate the weight of the organic fraction, as the more polar oxidized hydrocarbons are not extracted.

In view of this, a three-solvent sequential extraction with increasingly polar solvents was used for these samples. The sequential extraction is advantageous in that it allows separation of the organic materials into increasingly polar fractions in the initial stage of sample handling, as well as giving a more complete extraction of the organic materials.

The solvents chosen for this were cyclohexane, dichloromethane, and acetone with dielectric constants 2, 9, and 21, respectively. Cyclohexane extracts the nonpolar compounds that are present, i.e., aliphatic hydrocarbons, polycyclic aromatic hydrocarbons, and probably some nonpolar oxidized hydrocarbons. Dichloromethane (DCM) was chosen as the second solvent because it has been shown to be about 26% more efficient than benzene in removing organic compounds.[14] Infrared spectra of the DCM extracts indicate the presence of oxidized hydrocarbons in this fraction.

A final extraction with acetone was carried out because the amount of organic material extracted with this solvent relative to benzene has been shown to increase with increasing ozone concentrations.[14] Acetone would be expected to be particularly effective in removing highly polar oxidized hydrocarbons likely to be formed during pollution episodes. The acetone extracts were found to contain significant amounts of inorganic materials, including nitrate and trace metals, as well as organic compounds. A composite sample, analyzed for trace metals by atomic absorption and x-ray fluorescence, showed low levels of Cu ($0.06 \ \mu g/m^3$), K ($0.04 \ \mu g/m^3$), and Pb ($0.01 \ \mu g/m^3$). Infrared spectra of the acetone extracts indicated the presence of oxidized hydrocarbons and inorganic nitrate. On the basis of total nitrate levels measured during this period, the maximum amount of nitrate that could be present in the acetone extracts is approximately 23% by mass.

On the average, cyclohexane extracted approximately 2.5% ± 1.4% (S.D.)† of the total suspended particulates (TSP), while dichloromethane and acetone extracted 2.8% ± 1.2% (S.D.) and 6.4% ± 2.2% (S.D.), respectively. The total amount of solvent-extractable material found for the NYSAS-1976 samples, expressed as a percentage of total suspended particulates (11.7%) or in terms of $\mu g/m^3$ (13.3 $\mu g/m^3$), is very close to the yearly averages reported by Eisenbud and Kneip,[19] for particulate samples collected in 1968 and 1969 in

TABLE 1

CARBON-CONTAINING FRACTIONS OF THE AMBIENT AEROSOL, NYSAS–1976

Date	Extractable Organics *		Nonextractable Organics †	
	$\mu g/m^3$	%TSP	$\mu g/m^3$	%TSP
8/30–8/31/76	12.1	7.2	4.6	2.7
8/31–9/10/76	12.5	10.0	6.5	5.0
9/01–9/02/76	4.4	8.0	3.1	5.6

* Sum of material extracted with cyclohexane, dichloromethane, and acetone.
† Measured as elemental carbon.

Manhattan and extracted with benzene followed by acetone. No significant difference was found from season to season for the values reported for 1968 and 1969.[19]

Somewhat higher percentages of organic solvent (dichloromethane followed by dioxane)˙extractable material were reported by Miller and co-workers[20] for two 24-hour samples taken in August, 1972, in New York City: 21% and 26%. These two samples consisted of particulates ≤2 μm, suggesting that the organic materials present in suspended particulates are associated with smaller more respirable particles. The percentage of organic material present in New York City particulates is somewhat lower than that found in Los Angeles, where values as high as 43% have been reported during smog episodes.[21]

Daily variations in the concentrations of the organic solvent extracts and TSP are shown in FIGURE 1. As can be seen, there is some correlation between all three solvent extracts and TSP values.

Nonextractable carbon was determined for three of the samples which had been sequentially extracted. The values found are reported in TABLE 1 and

† S.D.—Standard deviation.

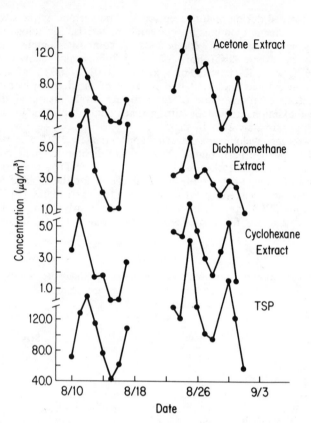

FIGURE 1. Aerosol concentrations of TSP and organic solvent extracts of TSP observed in New York City during NYSAS—1976.

are somewhat less than, but of the same order of magnitude as the values found for extractable organic compounds. The exact nature of the carbonaceous material remaining on the filter after organic solvent extraction is unknown. It is likely, however, that it consists of amorphous carbon as well as highly insoluble organic compounds.

Correlations of Aerosol Concentrations of Organic Solvent with Other Aerosol Components

While statistical correlations cannot be simply interpreted in terms of cause and effect, they often indicate unsuspected relationships and also suggest commonalities among variables. A nonparametric statistical test of correlation, the Kendall statistic,[17] was used to test pairwise dependence among the variables. Kendall's rank correlation coefficients and the level of significance were calculated for all possible pairs of data sets. Those variables found to correlate significantly, $p \leq 0.01$, are presented in TABLE 2.

Concentrations of dichloromethane- and acetone-extractable compounds from suspended particulates were found to significantly correlate with lead, zinc, cadmium, ozone, and traffic. Timewise variations of these solvent extracts and the trace metals are shown in Figures 2 through 4. In contrast, the cyclohexane-extractable compounds were not significantly correlated with any of these five variables.

Trace metals in urban aerosols generally originate from more than one source. Zinc, cadmium, and lead can originate from both automobiles and incinerators.[22] Kneip *et al.*[23] have estimated that in 1972 on an annual average basis, approximately 95% of the aerosol lead in New York City originated from automobiles. Increased use of lead-free and low-lead gasolines as well as seasonal variations in traffic may result in increased importance of incinerators to airborne lead concentrations, especially during summer months.

Ozone, on the other hand, is a secondary product resulting from photochemical reactions in the atmosphere. A seasonal pattern of ozone production,

TABLE 2

KENDALL RANK CORRELATIONS BETWEEN CONCENTRATIONS OF
ORGANIC SOLVENT EXTRACTS OF SUSPENDED PARTICULATES AND OTHER VARIABLES

Variables	n *	r †	p ‡
Dichloromethane Extract with:			
1. Ozone, Roosevelt Island, N.Y.	17	0.59	0.001
2. Pb	16	0.58	0.001
3. Traffic	16	0.59	0.001
4. Zn	13	0.58	0.003
5. Cd	16	0.45	0.007
Acetone Extract with:			
1. Pb	16	0.74	0.001
2. Zn	13	0.74	0.001
3. Cd	16	0.64	0.001
4. NO$_2$	8	0.91	0.001
5. Ozone, Roosevelt Island, N.Y.	17	0.45	0.006
6. Traffic	16	0.46	0.007
Cyclohexane Extract with:			
1. Mn	15	0.49	0.006
2. Fe	15	0.45	0.011
Trace Metals:			
1. Pb with Zn	13	0.74	0.001
2. Pb with Cd	16	0.61	0.001
3. Cd with Zn	13	0.85	0.001
4. Pb with Traffic	15	0.54	0.003
5. Pb with NO$_2$	7	0.78	0.008
6. Fe with Mn	16	0.82	0.001

* n is number of data pairs.
† r is Kendall's rank correlation coefficient.
‡ p is the probability that the observed correlation is not significant.

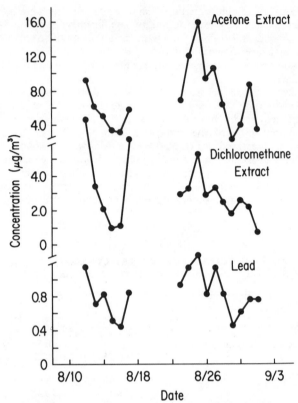

FIGURE 2. Aerosol concentrations of acetone and dichloromethane extracts of TSP and of lead observed in New York City during NYSAS—1976.

with maximum concentrations during the summer months, has been observed for this region of the United States.[24] Studies conducted in the New York Metropolitan area [24, 25] and in the eastern quadrant of the United States [26] indicate that long-range transport as well as local sources contribute to ozone concentrations.

Correlations of the dichloromethane- and acetone-extractable compounds with the daily maximum ozone concentrations at High Point, N.J., Roosevelt Island, N.Y., and Babylon, N.Y., during the same period are shown in FIGURE 5. The correlations for the New York summer aerosol are consistent with observations which have been made in Los Angeles. Grosjean and Friedlander [27] showed that hourly concentrations of the organic carbon aerosol fraction exhibited a morning and an afternoon maximum. The morning maximum in organic carbon concentration occurred prior to ozone formation and was attributed to direct emission and nonphotochemical gas to particle conversion of hydrocarbons and oxidized hydrocarbons from auto exhaust. The afternoon maximum was closely related to ozone concentrations and was attributed to particle formation via complex photochemical O_3-HC-NO_x interactions.

A multiple regression analysis was performed to test the hypothesis that emissions from automobiles, photochemical oxidation, and interactions between ozone and automotive emissions were the principle sources of oxidized organic compounds in suspended particulates in New York City.

An equation was set up with organic extractable mass as the dependent variable and O_3 concentrations, Pb concentrations, and a cross-product term that should represent the interactions of these sources, as independent variables. The results are summarized in TABLE 3.

The results showed that the highly polar compounds (those in the acetone fraction) were primarily associated with Pb concentrations and hence with automotive emissions. The dichloromethane fractions, which contain moderately polar compounds, correlated with the cross-product of O_3 and Pb, which indicates that the DCM extractable compounds were strongly influenced by interactions between O_3 and automotive emissions, presumably involving photochemical reactions. No significant regression results were obtained for the nonpolar compounds (cyclohexane extracts) with the preceding independent variables. Examination of the correlation coefficients in TABLE 2 and data in

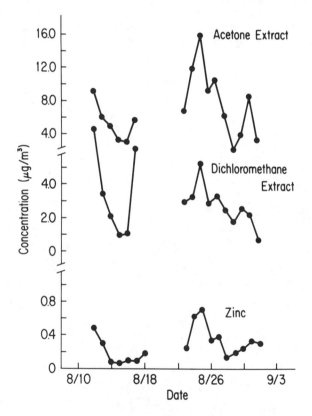

FIGURE 3. Aerosol concentrations of acetone and dichloromethane extracts of TSP and of zinc observed in New York City during NYSAS—1976.

TABLE 3

RESULTS OF MULTIPLE REGRESSION ANALYSIS

Solvent	Organic Compound Class	Significant Regression Variables (p <0.01)	F	% Variation Explained by Regression
Acetone	Polar	Pb	43.4	77
Dichloromethane	Slightly Polar	Pb X O₃	18.8	59
Cyclohexane	Nonpolar	—	—	—

FIGURE 6 indicates that the nonpolar compounds correlate strongly to particles of soil-like composition such as resuspended soil or fly ash from coal burning.

No significant correlations of extractable organic species ($p \leq 0.01$) were found to vanadium, sulfate, nitrate or ammonium ion or to minutes of sunlight per day. The average values of the concentrations of the three solvents grouped by wind direction were also examined, and no significant difference was found using Student's t-test. The data are, however, somewhat limited and future investigations of such relationships are planned.

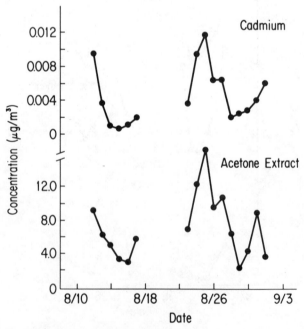

FIGURE 4. Aerosol concentrations of acetone extracts of TSP and cadmium observed in New York City during NYSAS—1976.

Polycyclic Aromatic Hydrocarbons (PAH)

The concentrations of several PAH compounds in the New York City ambient aerosol have been determined for some of the samples collected during NYSAS-1976. More extensive analyses are in progress. The ambient concentrations of benzo(a)pyrene, benzo(e)pyrene, benzo(ghi)perylene and chrysene/triphenylene‡ from August 10–17, 1976, and August 23–24, 1976, are shown

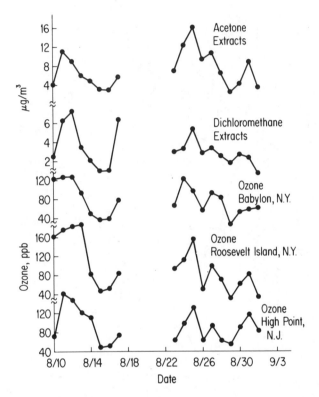

FIGURE 5. Aerosol concentrations of dichloromethane and acetone extracts and of ozone.

in FIGURE 7. The concentrations found are comparable to those which have been reported by Dong and co-workers [11] for samples collected in November and December, 1974, on the east side of Manhattan. The benzo(a)pyrene levels seem to be slightly lower than the average value reported (3.9 ng/m³) for samples collected from 1962–64 at Herald Square.[8] Herald Square is located

‡ Not separated by the analysis used.

near the center of the mid-Manhattan business district, about 0.5 miles to the west of the New York University Medical Center.

The correlations which were found, at a level of significance for which $p \leq 0.5$, are given in TABLE 4. For the data available for August 12–18, 1976, and August 23–24, 1976, the highest correlation coefficients are found for

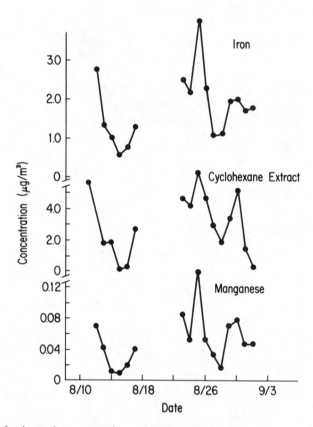

FIGURE 6. Aerosol concentrations of cyclohexane extracts of TSP and iron and manganese observed in New York City during NYSAS—1976.

benzo(a)pyrene with iron and manganese with the coefficient significant at greater than the 99% confidence level. The correlation coefficient found for benzo(a)pyrene with lead (0.67) in this study is close to that found by Colucci and Begeman (0.60).[8] Similar correlations were found in our 1976 study for chrysene/triphenylene with iron, manganese and lead. No significant correla-

TABLE 4

LINEAR CORRELATIONS BETWEEN SOME PAH COMPOUNDS AND TRACE METALS

Variables	Number of Pairs of Data	Pearson Correlation Coefficient
Benzo (a) pyrene with:		
Fe	7	0.92 *
Mn	7	0.84 *
Pb	7	0.67 †
Chrysene/Triphenylene with:		
Pb	7	0.81 *
Mn	7	0.77 *
Fe	7	0.74 †

* Significant correlation at the 99% confidence level.
† Significant correlation at the 95% confidence level.

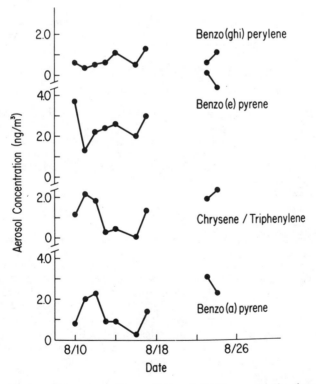

FIGURE 7. Aerosol concentrations of several PAH compounds observed during NYSAS—1976.

tion with vanadium was found for any of the PAH compounds measured during this time.

Bacterial Mutagenicity

Selected organic solvent extracts of total suspended particulate samples from NYSAS-1976 were assayed for direct mutagenic activity (no microsomal activation) using the Ames *Salmonella typhimurium* test system, strains TA 1535, TA 100, TA 1537, TA 1538, and TA 98. Additional testing with microsomal activation could not be carried out because of the limited sample size. The first two strains have been used to detect mutagens causing base-pair substitutions, while the remaining strains have been used to detect frameshift

TABLE 5

BACTERIAL MUTAGENICITY TESTING OF ORGANIC SOLVENT EXTRACTS
OF SUSPENDED PARTICULATE SAMPLES COLLECTED DURING NYSAS, 1976

| | | Revertants/μg * | |
| | Extraction | *S. typhimurium* | *S. typhimurium* |
Sample	Soivent	TA—98	TA—100
8/11–8/12/76	Dichloromethane	1.0	1.1
8/27–8/28/76	Dichloromethane	2.9	3.0
8/28–8/29/76	Dichloromethane	1.3	2.3
8/11–8/12/76	Acetone	0.65	— †
8/27–8/28/76	Acetone	0.36	1.2
8/28–8/29/76	Acetone	0.49	0.65
8/31–9/01/76	Cyclohexane	0.55	1.1
9/01–9/02/76	Cyclohexane	0.38	0.8

* (Revertants per plate for sample minus revertants per plate for blank) divided by μg of extract per plate.
† Insufficient material for testing.

mutagens. Strains TA 98 and TA 100 were found to be the most sensitive to the extracts tested.

All three solvent extracts, cyclohexans, dichloromethane, and acetone, exhibited mutagenic activity as shown in TABLE 5. The cyclohexane extracts are known to contain aliphatic hydrocarbons and polycyclic aromatic hydrocarbon compounds, none of which are direct acting mutagens. Many of the PAH compounds in this extract have been shown to exhibit bacterial mutagenic activity with the Ames test when the microsomal activation method is used. The activity observed here for the cyclohexane extracts indicates the presence of some unidentified direct acting mutagens as no microsomal activation was employed.

The dichloromethane extracts, which contain primarily oxidized hydrocarbons, exhibited the greatest activity, on the average, although the results, which are reported in terms of "revertants"/μg, are somewhat variable.

SUMMARY AND CONCLUSIONS

Twenty-four hour total suspended particulate (TSP) samples collected in New York City in two alternate weeks in August, 1976, were sequentially extracted with increasingly polar solvents—cyclohexane, dichloromethane, and acetone. On the average, cyclohexane extracted $2.5 \pm 1.2\%$ of the TSP, while dichloromethane and acetone extracted $2.8 \pm 1.2\%$ and $6.4 \pm 2.2\%$ of the TSP, respectively. The percentages and concentrations ($\mu g/m^3$) of organic material extracted from the samples were very similar to the 1968 and 1969 yearly averages reported for New York City by Eisenbud and Kneip.[19]

Correlations between the concentrations of the organic solvent extracts and trace metals, anions, ammonium ion, ozone, and sunlight and relationships to wind direction were investigated. The dichloromethane and acetone extracts, containing oxidized hydrocarbons, were found to be highly correlated with Pb, Zn and Cd, as well as O_3 and traffic. The correlations observed suggest automobiles, incinerators, and photochemical oxidation as important sources of oxidized hydrocarbons in New York City during the summer.

Cyclohexane extracts, containing nonpolar organic compounds, in contrast, were found to correlate with Fe and Mn concentrations, suggesting that these materials relate to soil-like particles.

Concentrations of several PAH compounds were determined for some of the samples. Benzo(a)pyrene concentrations were found to correlate most highly with concentrations of iron, manganese, and lead indicating sources such as resuspension, automobiles, and incinerators as contributing to aerosol burdens of this compound.

All three solvent extracts were found to exhibit mutagenic activity with the Ames test with no microsomal activation indicating the presence of unidentified direct-acting bacterial mutagens. The highest level of activity was found for the dichloromethane extracts.

ACKNOWLEDGMENTS

The authors wish to express their appreciation to a number of New York University Medical Center staff members: Robert Mallon for sample collection, James Miller and Bruce Naumann for trace metal analysis, Dr. David Bernstein and Raquel Pacheco for help with computer data analysis, Dr. Frank Mukai and Mrs. Irene Hawryluk for bacterial mutagenicity testing and special thanks to Dr. Theo. J. Kneip for many helpful discussions.

The work of New York University is supported by Grant No. RP439–1 of the Electric Power Research Institute and the American Petroleum Institute and is part of a Center program supported by Grant No. ES 00260, from the National Institute of Environmental Health Sciences, and Grant No. CA 13343, from the National Cancer Institute.

REFERENCES

1. Committee on Biologic Effects of Atmospheric Pollutants. 1972. National Academy of Sciences: Particulate Polycyclic Organic Matter. Printing and Publishing Office, National Academy of Sciences, Washington, D.C.

2. Committee on Medical and Biologic Effects of Environmental Pollutants. 1976. Vapor-Phase Organic Pollutants. Printing and Publishing Office, National Academy of Sciences, Washington, D.C.

3. HUEPER, W. C., P. KOTIN, E. C. TABOR, W. W. PAYNE, H. FALK & E. SAWICKI. 1962. Carcinogenic bioassays on air pollutants. Arch. Pathol. **74:** 89–116.

4. KOTIN, P., H. L. FALK & M. THOMAS. 1954. Aromatic hydrocarbons. II. Presence in the particulate phase of gasoline-engine exhausts and the carcinogenicity of exhaust extracts. A.M.A. Arch. Inc. Hyg. Occup. Med. **9:** 164–177.

5. WYNDER, E. L. & D. HOFFMANN. 1965. Some laboratory and epidemiological aspects of air pollution carcinogenesis. J. Air Pollut. Control Assoc. **15:** 155–159.

6. VAN DUUREN, B. L. 1972. Epoxides, hydroperoxides and peroxides in air pollution. Int. J. Environ. Anal. Chem. **1:** 233–241.

7. HOFFMANN, D. & E. L. WYNDER. 1968. Organic particulate pollutants. *In* Air Pollution. A. C. Stern, Ed. Vol. II. 2nd edit., pp. 215–247. Academic Press. New York.

8. COLUCCI, J. M. & C. R. BEGEMAN. 1971. Carcinogenic air pollutants in automotive traffic in New York. Environ. Sci. Technol. **5:** 145–150.

9. CUKOR, P., L. L. CIACCIO, E. W. LANNING & R. L. RUBINO. 1972. Some chemical and physical characteristics of organic fractions in airborne particulate matter. Environ. Sci. Technol. **6:** 633–637.

10. CIACCIO, L. L., R. L. RUBINO & J. FLORES. 1974. Composition of organic constituents in breathable airborne particulate matter near a highway. Environ. Sci. Technol. **8:** 935–942.

11. DONG, M., D. C. LOCKE & E. FERRAND. 1976. High pressure liquid chromotographic method for routine analysis of major parent polycyclic aromatic hydrocarbons in suspended particulate matter. Anal. Chem. **48:** 369–372.

12. AMES, B. N., J. MCCANN & E. YAMASAKI. 1975. Methods for detecting carcinogens and mutagens with the Salmonella/mammalian microsome mutagenicity test. Mutation Res. **31:** 347–364.

13. GORDON, R. J. 1974. Solvent selection in extraction of airborne particulate matter. Atmos. Environ. **8:** 189–191.

14. GROSJEAN, D. 1975. Solvent extraction and organic carbon determination in atmospheric particulate matter: The organic extraction—organic carbon analyzer (OE–OCA) technique. Anal. Chem. **47:** 797–805.

15. DAISEY, J. & M. A. LEYKO. 1979. Thin-layer/gas chromatographic method for the determination of polycyclic aromatic and aliphatic hydrocarbons in airborne particulate matter. Anal. Chem. **51:** 24–26.

16. EISENBUD, M. & T. J. KNEIP. 1975. Trace Metals in Urban Aerosols. EPRI Report No. RP–117. NTIS #Pb–248–324). Prepared for the Electric Power Research Institute, Palo Alto, Calif. October, 1975.

17. HOLLANDER, M. & D. A. WOLFE. 1973. Nonparametric Statistical Methods, pp. 185 ff. John Wiley & Sons. New York.

18. SNEDECOR, G. W. & W. G. COCHRAN. 1967. Statistical Methods, 6th edit. The Iowa State University Press. Ames, Iowa.

19. EISENBUD, M. & T. J. KNEIP. 1971. Trace Metals in the Atmosphere. Progress Report to the State of New York Departments of Health and Environmental Conservation, Contract No. C 20401, April 1, 1967–March 31, 1971.

20. MILLER, D. F., W. E. SCHWARTZ, J. L. GEMMA & A. LEVY. 1975. Haze Formation: Its Nature and Origin 1975, EPA Report No. 650/3–75–010. U.S. Environmental Protection Agency. Research Triangle Park, N.C. 27711.

21. HIDY, G. M., *et al.* 1974. Characterization of Aerosols in California, Vol. IV, p. 8–85. Final Report Submitted to the Air Resources Board, State of California, ARB Contract No. 358.

22. LAGERWERFF, J. V. & A. W. SPRECHT. 1971. Occurrence of environmental cadmium and zinc and their uptake inplants. *In* Trace Substances in Environmental Health, Proceedings of University of Missouri's 4th Annual Conference

on Trace Substances in Environmental Health. D. D. Hemphill, Ed. Vol. IV: 85. University of Missouri. Columbia, Missouri.

23. KNEIP, T. J., M. T. KLEINMAN & M. EISENBUD. 1975. Relative contribution of emission sources to the total airborne particulates in New York City. *In* Proceedings of the Third International Clean Air Congress. Düsseldorf, Germany.

24. BRUNTZ, S. M., W. S. CLEVELAND, T. E. GRAEDEL, B. KLEINER & J. L. WARNER. 1974. Ozone concentrations in New Jersey and New York: Statistical association with related variables. Science **186**: 257–259.

25. WOLFF, G. T., P. J. LIOY, R. E. MEYERS, R. T. CEDERWALL, G. D. WIGHT, R. E. PASCERI & R. S. TAYLOR. 1977. Anatomy of two ozone transport episodes in the Washington, D.C. to Boston, Massachusetts, corridor. Environ. Sci. Technol. **11**: 506–510.

26. WIGHT, G. D., G. T. WOLFF, P. J. LIOY, R. E. MEYERS & R. T. CEDERWALL. 1977. Formation and transport of ozone in the Northeast Quadrant of the U.S. *In* Proceedings of ASTM Conf. on Air Quality, Meteorology and Atmospheric Ozone. Boulder, Colorado.

27. GROSJEAN, D. & S. K. FRIEDLANDER. 1975. Gas-particle in the Los Angeles atmosphere. J. Air Pollut. Control Assoc. **25**: 1038–1044.

GASEOUS AND PARTICULATE HALOGENS IN
THE NEW YORK CITY ATMOSPHERE *

Kenneth A. Rahn, Randolph D. Borys,
Eric L. Butler, and Robert A. Duce

Graduate School of Oceanography
University of Rhode Island
Kingston, Rhode Island 02881

INTRODUCTION

For a number of years the University of Rhode Island has been studying the occurrence and distribution of the atmospheric halogens Cl, Br, and I in both polluted and remote areas. Sites that have been investigated include Providence (Rhode Island), Bermuda, Canada's Northwest Territories, Kansas, Arizona, Hawaii, American Samoa, and the South Pole. Recently, a new sampling system was developed and tested in a number of remote locations.[1] As a complement to these remote investigations, we decided to test the new system in the urban locations of Providence and New York City. This paper reports some of the initial data resulting from the New York study, with a brief comparison with the Providence data.

Our halogen sampling in New York began in July 1976. Coincidentally, at about this same time the large cooperative New York Summer Aerosol Study (NYSAS) of 1976 was also being organized, and so it was logical that the halogen study become a part of the larger study. Participating in NYSAS were the Institute of Environmental Medicine of New York University Medical Center, the Pierce Foundation of Yale University, the University of Rhode Island, the Environmental Measurements Laboratory of the U.S. Department of Energy, Brigham Young University, and the Interstate Sanitation Commission of New York, New Jersey, and Connecticut. NYSAS took place during August 1976; our halogen data of July and August 1976 are reported here for comparison.

SAMPLING

The basic simpling system used in this study has been described in detail elsewhere.[1, 2] It consists of a 0.4-μm Nuclepore™ filter 47 mm in diameter for particulate collection, followed by three 47-mm Whatman No. 41 cellulose filters impregnated with a 10% (W/W) solution of tetrabutyl-ammonium hydroxide (TBAH) in a 10% (V/V) glycerol-water mixture (for collection of inorganic gaseous halogens), followed by six 0.5-g beds of 10- to 12-mesh activated charcoal for collection of organic halogen gases. All four filters were placed in a single 47-mm filter holder, with the Nuclepore prefilter separated

* Presented at the Aerosol Technology and Air Pollution Session, The May 1977 Industrial Hygiene Conference, American Industrial Hygiene Association, New Orleans, Louisiana, May 26, 1977.

0077–8923/79/0322–0143 $01.75/0 © 1979, NYAS

from the others by a nylon spacer. The filter holder was mounted over a glass tube containing the charcoal. Air flow was provided by a carbon-vane vacuum pump. Typical initial flow rates were 17 to 23 1 min⁻¹.

Nuclepore filters plug up rapidly during sampling. In the polluted air of New York City this effect was sometimes severe. Generally, the Nuclepore prefilters had to be changed during the basic weekly sampling period, when the flow rates had decreased to well below half the initial value. The worst case occurred during week 2, when six filters were required. The impregnated filters and activated charcoal had capacity enough for a week's sampling, however, and so were only changed at the beginning of each week.

During weeks 6 and 7, an Andersen 7-stage, 28 1 min⁻¹ (1 cfm) cascade impactor was placed ahead of the Nuclepore prefilter, in order to provide some information about the variation of the halogens' mass with particle size. Thin polyethylene films were placed on the regular aluminum impaction plates and served as impaction surfaces. The films had very low halogen impurity levels, and were analyzed in the same fashion as were the Nuclepore filters. During these runs the Nuclepore filters plugged up much less rapidly than they did without the impactor, and average flow rates of 17 1 min⁻¹ (0.6 cfm) were obtained. The 50% cutoff diameters for the Andersen impactor under these conditions are shown in TABLE 1.

Sampling took place on the roof of the New York University Medical Center Residence Hall, a 14-story building on the east side of midtown Manhattan. Samples were generally begun on Thursdays or Fridays during July and Mondays or Tuesdays during August. The sampling record is given in TABLE 2.

ANALYSIS

The samples of this study were analyzed by neutron activation methods developed at URI, which have been described elsewhere.[1, 2] Briefly, the samples were irradiated in the nuclear reactor of the Rhode Island Nuclear Science Center for 2 to 20 minutes, allowed to decay for 6 to 15 minutes, then counted nondestructively with a Ge(Li) gamma-ray detector coupled to a 4096-channel analyzer. Standard halogen mixtures spotted onto cellulose filters were run according to the same scheme and compared to the samples by means

TABLE 1

50% EQUIVALENT CUTOFF DIAMETERS OF THE ANDERSEN IMPACTOR

	ECD, μm	
Stage	28 ℓ min⁻¹	17 ℓ min⁻¹
1	9.2	11.9
2	5.5	7.1
3	3.3	4.3
4	2.1	2.7
5	1.1	1.42
6	0.65	0.84
7	0.43	0.55

TABLE 2

SAMPLING DATA FOR HALOGENS IN NEW YORK CITY

Sample	Dates	Volume, SCM	Type of Sample *
NYC HAL 1	8–15 July 1976	82.0	Nuclepore filter
NYC HAL 2	15–22 July 1976	64.5	6 sequential Nuclepore filters
NYC HAL 3	23–29 July 1976	68.1	3 sequential Nuclepore filters
NYC HAL 4	29–31 July 1976	39.1	Nuclepore filter
NYC HAL 5	2–5 August 1976	63.0	Nuclepore filter
NYC HAL 6 (AS 1)	10–16 August 1976	146.8	Andersen impactor plus Nuclepore afterfilter
NYC HAL 7 (AS 2)	17–23 August 1976	134.5	Andersen impactor plus Nuclepore afterfilter
NYC HAL 8	23–26 August 1976	39.2	2 sequential Nuclepore filters

* Filter or impactor always followed by impregnated filters and activated charcoal.

of an Al flux monitor which was always co-irradiated. Occasionally chemical separation schemes had to be used for I in these urban samples.[2] Separations were always done after irradiation.

RESULTS AND DISCUSSION

The analytic results for the first eight samples during July and August 1976 are given in TABLE 3 as ng/SCM. Several interesting features of this table can be immediately seen.

First, the different components showed different trends with time. The most pronounced trend was shown by particulate Cl, probably because this component was derived the most directly from sea-salt aerosol, as will be discussed below. Particulate Cl decreased smoothly from 780 ng/SCM during week 1 to 104 ng/SCM during week 4, then increased smoothly back up to 640 ng/SCM during week 8. Particulate I showed some indication of a parallel trend, but particulate Br showed nearly no trend at all. No significant trends were shown by any of the gaseous halogens.

Distinct differences in variability of concentration from sample to sample were seen for the nine halogen components, as shown by the coefficients of variation listed in this table. In general, the particulate halogens were about as variable as the inorganic gaseous halogens (40 to 50%, except for inorganic gaseous Cl), but the organic gaseous halogens were significantly less variable (20 to 25%, except for I). We believe, but are not sure, that the variability of the organic gaseous I was an artifact of its notoriously high variability in the activated charcoal, at least partly because its extreme variability here seemed to follow no pattern at all. If so, the generalized pattern of organic gaseous halogens being less variable temporally than either particulate or inorganic halogens would be the analogue of what was seen earlier for spatial variations over North America.[1] These spatial variation patterns were interpreted in terms of longer residence times for the organic gaseous halogens, which seems to make sense because of the generally lower reactivity of organic compounds such as halogenated hydrocarbons relative to simple inorganic halogens gases

TABLE 3

New York City Halogen Data (ng/SCM)

Sample	Particulate			Inorganic Gas			Organic Gas		
	Cl	Br	I	Cl	Br	I	Cl	Br	I
1	780±40	167±4	<1.5	11,800±200	117±3	7.3±1.0	23,000±1000	150±8	1.8±1.5
2	620±120	260±10	5.1±0.9	13,900±100	200±2	12.7±0.6	23,000±1000	210±10	12.0±2.3
3	370±20	142±4	4.0±0.5	2,900±100	94±3	4.8±0.3	18,000±500	113±9	5.9±2.4
4	104±10	230±10	2.3±0.4	5,600±100	117±3	14.2±0.5	16,100±200	114±9	<5
5	300±30	19.8±3.9	2.8±1.1	5,300±100	210±10	12.1±0.4	23,000±1000	170±7	3.3±1.6
6	380±30	171±5	2.2±1.5	1,980±20	150±1	9.9±0.2	16,300±100	202±4	15.5±1.3
7	590±20	210±10	2.4±0.4	2,200±100	178±1	7.3±0.2	12,400±100	164±5	3.1±0.7
8	640±40	330±10	5.7±1.4	5,400±100	370±10	19.1±0.6	22,000±1000	200±20	5.5±2.6
Mean±σ	470±230	191±92	3.5±1.4	6,140±4,420	187±84	10.9±4.6	19,200±4100	165±38	6.7±5.1
Coefficient of Variation	(49%)	(48%)	(40%)	(72%)	(45%)	(42%)	(21%)	(23%)	(76%)
Gas/Particulate Ratios	1	1	1	13.1	0.98	3.1	41	0.86	1.91

such as HCl, etc., which are both water-soluble and highly reactive. Particulates should also have shorter residence times than organic gases, because the former are readily scavenged by precipitation, and can fall out or impact on surfaces. Long-residence-time components ought to build up a higher atmospheric background concentration than would short-residence-time components, and so would be expected to be less variable in both time and space, the behavior that is apparently seen here.

Within the particulate halogens, I had a lower variability than either Cl or Br; it will be pointed out below that it is also associated with the smallest particle-size aerosol. Inorganic gaseous Cl was much more variable than either inorganic gaseous Br or I, which suggests that there may be some strong local source for this form of Cl. Incineration of chlorine-containing plastics such as

TABLE 4

TYPICAL GASEOUS AND PARTICULATE HALOGEN CONCENTRATIONS
FROM VARIOUS AREAS
(all data ng m^{-3} STP)

Locations	No. Samples	Particulate			Inorganic Gas			Organic Gas		
		Cl	Br	I	Cl	Br	I	Cl	Br	I
Northwest Territories	5	22	0.4	0.2	80	0.4	0.4	1,200	14	3
Kansas	3	33	11	2.6	170	8	5	1,400	39	12
Arizona	7	66	3.3	1.3	570	11	11	680	17	5
Bermuda	1	640	13	3.8	380	24	17	1,000	68	28
New York City	8	470	191	3.5	6140	187	11	19,200	165	7
$\sigma_{NYC}(\%)=$		(49)	(48)	(40)	(72)	(45)	(42)	(21)	(23)	(76)
Gas/particulate ratios		1	1	1	13	1	3	40	0.9	2
NYC/Remote		10	100	3	20	40	3	20	8	1

PVC may be a possibility. Finally, as noted above, the strangely high variability of organic gaseous I compared to Cl or Br, for which there is no obvious reason, may be an artifact of the activated charcoal.

The bottom line of TABLE 3 gives the gas/particulate ratios for each of the halogens. Cl occurs overwhelmingly as a gas ($>98\%$), while Br (65%) and I (83%) are much less so but still primarily gases. As shown below, the predominance of gaseous over particulate halogens is a worldwide phenomenon.

In TABLE 4 the New York halogen results are compared with typical data determined at URI for halogens in various remote areas. In the lowest line of this table the ratio of the concentration in New York to that in a "typical" remote area is given for each of the nine halogen components. In general, this ratio is largest for the particulate halogens, intermediate for the inorganic gaseous halogens, and smallest for the organic gaseous halogens. Thus, the "polluting order" for the halogen forms is particulate $>$ inorganic gas $>$ organic gas. When the various elements are considered, Br seems to have the greatest ratios, hence is the greatest pollutant, Cl is intermediate, and I is the least polluting in cities. In fact, I with its three NYC/remote ratios of 3, 3, and 1 is hardly a pollutant at all. It is perhaps the first trace element we have found in New York City to behave in this way.

TABLE 5 gives a summary of the data from the two impactor runs concerning the variation of the halogens' mass with particle size. In general, the two runs gave very similar results, perhaps because they were taken during consecutive weeks, but also because the preferred particle size of an element in the aerosol seems not to vary nearly as widely as does its concentration. Cl was associated with the giant particles (d $>$ 2 μm), and had mass-median diameters (MMDs) of 6 μm for both runs. This large particle size is undoubtedly because its primary local source was sea-salt particles. In fact Na, not shown here, also had its mass primarily associated with the giant particles,[3]

TABLE 5

HALOGEN CONCENTRATIONS VS. PARTICLE SIZE

Andersen Stage	Cl (ng/SCM)		Br (ng/SCM)		I (ng/SCM)	
	AS1	AS2	AS1	AS2	AS1	AS2
1	96	176	11.2	13.9	—	0.31
2	74	132	12.6	14.7	—	<0.5
3	66	71	14.5	16.0	—	0.20
4	61	78	24	26	—	<0.2
5	27	60	24	34	—	0.15
6	19	17	12.5	11.2	—	0.24
7	7	3	15.0	13.2	—	0.34
8	<30	56	57	85	—	0.78
Total	350	590	171	210	—	2.4
MMD, μm	6	6	1.5	1.5	—	0.6

as would be expected for sea salt. Its concentration in the giant range was roughly equal to the Cl concentration, as is typical for marine-derived aerosol.[4] Br was found on both giant and large (0.2 μm $<$ d $<$ 2 μm) particles, but primarily with the large particles. Its MMD was 1.5 μm for both samples. The giant-particle Br was probably from sea salt, and the large-particle Br was probably derived from automobile emissions. And finally, I was strongly associated with the finest particles, having a MMD of only 0.6 μm in the second sample. In the first sample it was below the detection limit. This fine-particle I was probably derived indirectly from sea salt, because even over the open ocean where the sea is surely the main source, I is associated with these fine particles.[5]

It is interesting to note that these particle-size results are completely in accord with previous observations for the halogens, none of which were, however, from New York City. Rahn[4] compiled the available data on elemental particle sizes in the aerosols of various regions, and gave as median MMDs,

4 μm for Cl, 0.9 μm for Br, and 0.6 μm for I. Note how close these values are to the New York City values.

New York City was the first city where we used this new collection system for halogens. During August 1976, however, four nearly parallel samples were taken in Providence, Rhode Island for comparison. The results of this study will be written up in detail elsewhere; for our purposes here it is sufficient to compare summaries of the New York and Providence data, as presented in TABLE 6, which will now be described.

Concentrations at the two locations were basically similar. The three forms of I were each about twice as concentrated in New York as in Providence, perhaps because New York is nearer the sea. The particulate levels of Cl were about the same, but the inorganic gaseous level was three times higher in New York and the organic gaseous level was five times higher in Providence. This latter concentration of over 100 μm/SCM is very high in absolute terms, and represents an anomalous source whose identity is not yet known. For Br, the particulate and inorganic gaseous forms were higher in Providence, perhaps because the Providence site was at ground level and thus closer to traffic and the release of Br from leaded gasolines, but the organic gaseous level was about the same. All in all, then, there were few fundamental differences between halogen abundances in New York and Providence.

Concerning the coefficients of variation, the particulates were generally more variable in New York, the inorganic gases were roughly the same in both locations, and the organic halogen gases showed no trend. There was also little pattern by element, except for organic Cl, which was much more variable in Providence (presumably because of a single anomalous source for its high concentrations).

The gas/particulate ratios were higher for Cl and lower for Br in Providence, but the same for I in both locations. In Providence 99.6% of the Cl was gaseous, 84% of I was gaseous, but only 39% of Br was gaseous.

SUMMARY

A new sampling system that segregates atmospheric halogens into particulate, inorganic gaseous, and organic gaseous components was used for study of the New York City aerosol during July and August 1976. In New York City (and Providence) the majority of all three atmospheric halogens was gaseous rather than particulate. Cl and I were found primarily as organic gases, while Br was primarily inorganic gaseous. Cl tended to be more gaseous in Providence, Br less gaseous, and I about equally gaseous in Providence and New York.

Particulate Cl showed a smooth temporal trend in New York, but the other eight halogen components showed little or no trend. As a rule the organic halogen gases varied less than did the particulate or inorganic gaseous components, implying that the organic gases had the longest atmospheric residence times. This agreed with spatial-variation tendencies reported earlier.[1] I was, in general, less variable, hence probably longer-lived, than Cl or Br.

In comparing New York with remote areas, the order of enrichment or pollution in New York is particulates > inorganic gases > organic gases. The order by element is Br > Cl > I. I has nearly the same abundance in New York as it has in remote areas.

TABLE 6

ATMOSPHERIC HALOGENS IN NEW YORK CITY AND PROVIDENCE

Location	Particulate			Inorganic Gas			Organic Gas		
	Cl	Br	I	Cl	Br	I	Cl	Br	I
				Concentrations, ng/SCM					
New York	470±230	191±92	3.5±1.4	6140±4420	187±84	10.9±4.6	19,200±4100	165±38	6.7±5.1
Providence	450±80	660±180	1.7±0.4	1700±500	260±90	6.0±3.0	103,000±68,000	168±66	3.2±2.1
				Coefficients of Variation, %					
New York	49	48	40	72	45	42	21	23	76
Providence	18	27	23	28	34	50	65	39	66
				Gas/Particulate Ratios					
New York	1	1	1	13.1	0.98	3.1	41	0.86	1.91
Providence	1	1	1	3.8	0.39	3.5	230	0.25	1.9

Within the aerosol Cl is associated with large particle sizes, Br is associated with intermediate particle sizes, and I is found on small particle sizes. These data also agree with trends established in other parts of the world.

ACKNOWLEDGMENTS

We wish to acknowledge the generous access to the sampling site in New York City and the enthusiastic help provided by Dr. Theo J. Kneip of the New York University Environmental Medical Center throughout this study. Mr. Robert Mallon, the stationkeeper during July and August 1976, faithfully and cheerfully changed many samples for us. All samples were irradiated and counted using facilities of the Rhode Island Nuclear Science Center, Narragansett, Rhode Island. This work was supported by NSF Grant ATM 75–23725, "Atmospheric Chemistry of the Halogens—Natural and Anthropogenic."

REFERENCES

1. RAHN, K. A., R. D. BORYS & R. A. DUCE. 1976. Tropospheric halogen gases: inorganic and organic components. Science **192:** 549–550.
2. RAHN, K. A., R. D. BORYS & R. A. DUCE. 1977. Special Environmental Report No. 10. Air pollution measurement techniques. World Meteorological Organization Report No. 460. Geneva, Switzerland, pp. 172–178.
3. BERNSTEIN, D. M. & K. A. RAHN. 1979. New York Summer Aerosol Study: Trace element concentrations as a function of particle size. Ann. N.Y. Acad. Sci. **322.** (This volume.)
4. RAHN, K. A. 1976. The chemical composition of the atmospheric areosol. University of Rhode Island Technical Report, 1 July 1976, 265 pp.
5. DUCE, R. A., J. W. WINCHESTER & T. W. VAN NAHL. 1965. Iodine, bromine, and chlorine in the Hawaiian marine atmosphere. J. Geophys. Res. **70:** 1775–1799.

A DISCUSSION
OF THE NEW YORK SUMMER AEROSOL STUDY, 1976

Paul J. Lioy * and George T. Wolff *

Interstate Sanitation Commission
New York, New York 10019

Brian P. Leaderer

Department of Epidemiology and Public Health
John B. Pierce Foundation Laboratory
Yale University School of Medicine
New Haven, Connecticut 06519

INTRODUCTION

The New York Summer Aerosol Study of 1976 (NYSAS) was designed to obtain information on the nature of the New York summer aerosol in the New York metropolitan area. Detailed information on the study objectives, organization, equipment, aerosol and meteorologic parameters measured and findings in each of the subject areas investigated in the study were presented in the preceding papers. This paper draws upon the results and discussions in the individual subject papers to indicate the interrelationships in the physical and chemical characteristics of the summer aerosol measured at the New York City site, and the probable causes for these relationships. Long range transport was found to have a major influence on the levels of sulfate ($SO_4^=$) and total suspended particulates (TSP) in the New York area.[1] Because of the importance of these findings to the development of effective control strategies, all NYSAS data have been examined as a whole in order to clarify the significance of the transport phenomenon and to confirm the extent of particulate transport.

PHYSICAL AND CHEMICAL PROPERTIES

Continuous hourly size distribution measurements reported by Knutson et al.[2] showed the diurnal pattern of particles less than 0.1 μm (freshly generated aerosol) to closely follow the diurnal traffic pattern and to a lesser extent the summer diurnal power demand pattern, suggesting automotive sources to be the primary source of Aitken nuclei in New York during the summer. The diurnal pattern for particles less than 0.1 μm remained constant when analyzed for days of varying relative humidity and wind direction. The diurnal pattern for the volume of particles between 0.1 μm and 1.3 μm ($V_{0.1-1.3}$), the size range that accounts for visibility degradation, remained surprisingly flat even when

* Current affiliations:

Paul J. Lioy
Institute of Environmental Medicine
New York University Medical Center
New York, New York 10016

George T. Wolff
Environmental Science Department
General Motors Research Laboratories
Warren, Michigan 48090

0077-8923/79/0322-0153 $01.75/0 © 1979, NYAS

analyzed for days of varying relative humidity and wind direction and could not be related to the diurnal traffic or power demand pattern. The diurnal pattern for light scattering, b_{scat}, reported by Leaderer et al.[3] was consistent with the pattern for $V_{0.1-1.3}$.

The lack of an afternoon peak in the $V_{0.1-1.3}$ and b_{scat} in the summer diurnal patterns, when photochemical activity would be maximal, suggests that photochemically generated aerosol in New York City in the light scattering size range resulting from local pollutant emissions is a relatively small component in relation to the aged aerosol within the ambient air. This finding is in direct contrast to results obtained in the ACHEX study conducted in California where the diurnal pattern recorded for $V_{0.1-1.0}$ clearly showed an afternoon peak corresponding to a photochemically generated aerosol resulting from local pollutant emissions.[4, 5] The difference between diurnal $V_{0.1-1.0}$, b_{scat} (for New York only) and total number patterns for New York and Los Angeles can be clearly seen in FIGURE 1. While the total number diurnal patterns are somewhat similar in shape for both locations, the diurnal patterns for $V_{0.1-1.0}$ are markedly different. The existence of a photochemical aerosol generated in Los Angeles in the 0.1 μm to 1.0 μm size range is clearly detectable whereas the New York curve suggests the absence of such activity. During the course of the New York study, there were no data collected which would allow for an hourly or even daily evaluation of atmospheric stability in New York City. The unavailability of such information prevented a normalization of the diurnal patterns for $V_{0.1-1.3}$ and b_{scat} based on atmospheric stability. Such a normalization could conceivably alter the diurnal patterns and show the local production of some

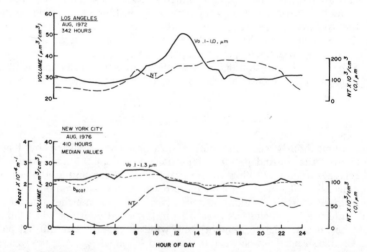

FIGURE 1. The diurnal pattern for total number of particles and volume of particles between 0.1 μm–1.3 μm for New York and Los Angeles and the diurnal pattern for light scattering (b_{scat}) for New York. Los Angeles data were obtained from the ACHEX study (August 1972)[5] and represents average hourly values determined from a total of 342 size distribution measurements. The New York total number $V_{0.1-1.3}$ curves were developed from 410 size distribution measurements taken during the NYSAS and presented as median values. The diurnal b_{scat} curve was developed from over 1200 hourly average measurements taken during the NYSAS in New York.

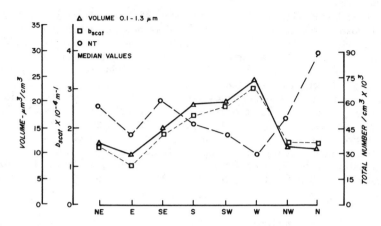

FIGURE 2. Median hourly readings of $V_{0.1-1.3}$, total number of particles and b_{scat} vs. eight wind direction ranges.

afternoon photochemical aerosol in New York that is masked by higher dispersion conditions in the early afternoon compared to the remainder of the day.

The meteorologic variable found to have the most pronounced effect on hourly size distribution and light scattering measurement in New York City during the summer was wind direction. FIGURE 2 shows that the concentration of total number of particles (essentially those less than 0.1 μm) was greatest when wind directions were from the north, northeast, northwest, and southeast and lowest with wind directions from the south, southwest, and west. Winds from the south, southwest, and west were found to result in high hourly readings of $V_{0.1-1.3}$ and b_{scat}, while other wind directions were associated with lower readings. Relative humidity had no effect on the small particles (less than 0.1 μm) while light scattering and $V_{0.1-1.3}$ was found to be positively correlated with relative humidity.

Approximately 51% of the mass was reported by Lippmann *et al.*[6] to be in the submicron size range, and of this, the largest chemical component was sulfate. Tanner *et al.*[7] found the predominant cation associated with the sulfate to be ammonium ($NH_4^+/SO_4^{-2} \simeq 0.75$) with over 85% of the total sulfate less than 2.0 μm in size. Nitrate concentration reported by Kleinman *et al.*[8] made up less than 2% of the submicron mass, while the analyses of Daisey *et al.*[9] found that less than 7% of the submicron aerosol was organic in nature, principally the oxidized hydrocarbon fraction. Nitrate levels in New York were only 20% of those found in Los Angeles and organics were only 50% of Los Angeles levels.

A strong systematic relationship was observed in New York City during the course of the study between visibility, measured by the extinction coefficient for scattered light (b_{scat}), oxidant concentrations (as measured by ozone), and sulfates. The observed relationship, shown in FIGURE 3, suggests that decreasing average daily visibility in New York during the summer is strongly associated with increasing average daily sulfate levels ($r = 0.90$, $p < 0.001$), and less strongly associated with maximum ozone concentrations ($r = 0.67$, $p < 0.001$). The strong relationship between visibility degradation and sulfates cannot be readily explained, since the other components of the <1.3 μm particles, which should also be related to light scattering, weakened the correlation. The association between sulfate concentration and nephelometer readings was much closer than that between accumulation mode mass and nephelometer readings.[6]

The strong relationship between ozone, sulfates, light scattering (visibility), TSP and between ozone and the oxidized hydrocarbon component of the TSP observed during the summer study could indicate the presence of a locally generated photochemical aerosol (principally sulfates) in New York City during the summer resulting from local primary pollutant emission. However, this observation is not consistent with the flat diurnal pattern recorded for $V_{0.1-1.3}$ and b_{scat}, which suggests little or no locally generated photochemical aerosol. Examination of the diurnal pattern for $V_{0.1-1.3}$ and b_{scat} on days of high ozone (greater than 80 ppb) showed that, while $V_{0.1-1.3\ \mu m}$ and b_{scat} were elevated,

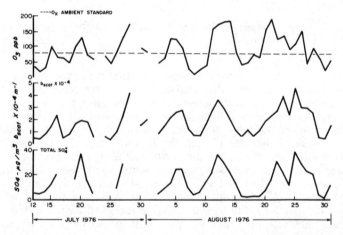

FIGURE 3. Daily 24-hour values of sulfates and light extinction (b_{scat}) and daily maximum hourly ozone concentration recorded in New York during July and August 1976.

there was no detectable maximum occurring during the early afternoon, which would be indicative of local photochemical aerosol generation. In addition, no significant diurnal variations in submicron sulfate and nitrate concentrations or composition were found further indicating the absence of a significant fraction of local photochemically generated aerosol in New York City. In fact, it was observed that the size and chemical nature of the sulfate seen in New York City, even during periods of high ozone, was very similar to data obtained in rural Illinois,[7] supporting the hypothesis that the submicron sulfate aerosol in New York is an aged aerosol that may have been formed in hazy air masses from precursors emitted upwind of New York City rather than formed as a result of local photochemical activity. The build-up of ozone concentrations, TSP, sulfates and increased light scattering observed during the study was found by Lippmann et al.[6] to be directly related to anticyclonic activity.

There was a less extensive body of data available to fully evaluate the variations in heavy metals and organic fractions of the TSP in the New York summer aerosol, although, some interesting relationships were observed. The oxidized hydrocarbon fractions obtained from 24-hour TSP samples were found to be strongly correlated with ozone, lead, zinc, cadmium, and traffic volume on the FDR Drive suggesting automobiles, incinerators and photochemical oxidation as important sources of oxidized hydrocarbons in New York City during the summer. Short-term variations in the concentrations of oxidized hydrocarbon fraction of particulate were not determined. Consequently, it was not possible to determine if local photochemical activity in New York City, during periods of high ozone, was principally responsible for oxidized hydrocarbons or if direct emissions of oxidized hydrocarbons from automobiles and incinerators, which were allowed to accumulate under the influence of stagnant air masses high in ozone, were responsible for elevated oxidized hydrocarbons. There is also the possibility that a portion of the oxidized hydrocarbons could be accounted for by the background levels in the regional air mass.

The oxidized hydrocarbon fractions have been found to be biologically active,[9] as determined by the Ames test,[10] and may be of significance to human health. Further investigation of the sources of these materials and clarification of the possible role of transport would thus be appropriate. Heavy metal concentrations measured during the course of the study and reported by Bernstein *et al.*[11] did not exhibit any clear patterns but were observed to be highest during periods of anticyclonic activity.

Daily sulfates, light scattering, and daily maximum hourly ozone concentrations measured at the New York City site during July and August, 1976, were averaged by daily prevailing wind direction and are shown in FIGURE 4. Daily average sulfates were found to be highest with wind directions from the southwest with substantial levels recorded when prevailing daily winds from the southeast, south, and west. Other daily prevailing wind directions (north, northeast, and northwest) were associated with lower sulfate concentrations. Daily average light scattering readings and maximum daily ozone concentrations exhibited a wind directional pattern similar to that for sulfates except that the southeasterly wind direction resulted in a more pronounced effect in increasing light scattering (reduced visibility) and westerly winds resulted in slightly higher ozone concentrations. There were insufficient data available to examine the effect of varying daily wind directions on levels of the organic fractions of the TSP and heavy metal components.

It appears that in New York City during the summer, particulates less than 0.1 μm are strongly related to automotive sources and highest during northerly surface wind direction, while the visibility-reducing aerosol (0.1–1.0 μm) is principally made up of an aged aerosol (primarily sulfates) and associated with regional visibility-reducing stagnant air masses high in ozone and arriving with surface winds of southeast, south, southwest, or west directions with the highest levels recorded on southwest and westerly wind directions. Associated with these stagnant hazy air masses rich in ozone is a build-up of the oxidized hydrocarbon component of the TSP and various heavy metals.

Transport Phenomenon

In addition to the examination of the New York City[12] aerosol, the New York Summer Aerosol Study (NYSAS) included analyses of aerosols collected at High Point, New Jersey.[13] During August 1976, fourteen coincident measurements of total suspended particulates (TSP) and sulfates ($SO_4^=$) were made daily. Subsequently, from the particulate data, backward trajectory analyses, and particle size distributions, it was determined that as much as 73% of the $SO_4^=$ and 35% of the TSP were transported through High Point to New York City. (This was a conservative estimate since only particles

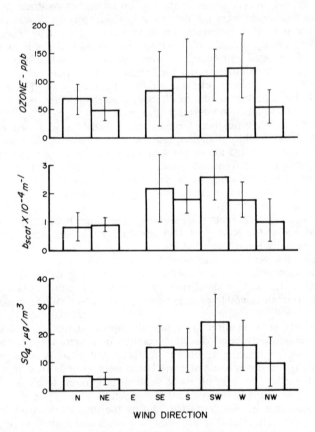

FIGURE 4. Average daily readings of sulfates and light scattering (b_{scat}) and daily maximum hourly ozone concentration vs. prevailing daily wind direction for July and August 1976 in New York.

<2.0 μm in diameter were considered transportable from High Point to New York City and it was assumed that mass fractionation percentages were identical for both sites in the settling calculations).

As with ozone, the $SO_4^=$ apparently was produced in the photochemical smog complex associated with high pressure systems in route from the midwestern to the eastern United States.[14] FIGURE 5 illustrates the relationship between $SO_4^=$ and wind direction. In each case, the high (>7.5 μg/m³) sulfate concentrations were recorded on days with a westerly wind trajectory, which is normally associated with return flow around the high pressure system.

The relationship between $SO_4^=$ and wind trajectories showed that in most cases, westerly winds through the mixing layer (SW on surface) yielded high sulfate concentrations in New York City. Southwesterly winds, which are frequent in this region during the summer, are associated with (1) high fine particle concentrations, (2) increased b_{scat} values, (3) increased ozone in the northeast corridor, and (4) high temperatures and relative humidity at New York City,[2, 3, 15, 16] High Point,[12, 15] and New Haven as noted in the preceeding discussion. West to southwest winds in the northeastern United States are generally associated with the return flow around high pressure systems. Therefore, concentrations of $SO_4^=$ above 7.5 $\mu g/m^3$ should exist along the northeastern coast of the United States in conjunction with the return flow sector of a high pressure system. This is illustrated in TABLE 1 by a comparison of the average backward wind trajectory terminating in New York City for each 24-hour sampling period in August 1976 (thereby including the directional influence of an entire day) with the actual $SO_4^=$ values, using 7.5 $\mu g/m^3$ as the high-low cutoff. The association of high pollution readings with the return flow was also seen in the pollution forecast completed for the NYSAS by Lippmann *et al.*[6] and is partly reproduced in TABLE 1. An example of the persistence of these episodes occurred between August 25th (noon) through August 29th (noon) when the highest $SO_4^=$ concentration in the NYSAS was recorded, and the average concentration exceeded 29 $\mu g/m^3$. In this case, a fresh (low $SO_4^=$) air mass entered the region on 8/24/76 and was assimilated into the return flow of an older high pollution air mass, that had been in the area for eight days. During the same period very high concentrations of accumulation mode particles,[2] and high b_{scat} values were recorded.[3] Similar concentration trends were observed for $SO_4^=$, TSP, ozone, and conjunctive low visibility on the backside of all the high pressure systems that affected the area during the study.

In the NYSAS, diurnal b_{scat} measurements consistently showed a lack of any diurnal variation.[3] The urban area historically has been described as a source that has a diurnal pattern for aerosols which reflects traffic patterns and commercial and industrial activity. On the basis of such a pattern, local pollution would be expected to show rapid increases in the morning, a peak during the day, and gradual decreases at night. The lack of a pronounced b_{scat} daily peak in the New York area indicates that the light scattering particulates originate upwind of New York City.

Lippmann[6] correlated b_{scat} with Hi-Vol sampler mass concentrations of $SO_4^=$, TSP, and accumulation mode suspended particulates (AMSP) in New York City during the NYSAS and found that the best correlation was between b_{scat} and $SO_4^=$ ($r = 0.85$). Leaderer *et al.*,[3] using the Brookhaven diffusion sampler $SO_4^=$ data, found that the 0.03 to 5 μm $SO_4^=$ mass correlated with the nephelometer with an $r = 0.94$. On the basis of these results, it appears that the nephelometer could be used as an indicator of sulfates. Furthermore, the flat b_{scat} diurnal pattern supports the concept of transported sulfate.[6]

Kleinman *et al.*[17] have employed the technique of dispersion normalization in the analysis of ambient particulate data to compensate for the effects of pollutants; thus any masked diurnal cycle is revealed. Using this technique for the $SO_4^=$ detected in the metropolitan area, no afternoon peak was evident, again suggesting that the source of the $SO_4^=$ was not local.

FIGURE 5. Trajectories associated with coincident $SO_4^=$ sampling at High Point, New Jersey and New York City, New York. (a) high sulfate days > 7.5 $\mu g/m^3$; (b) low sulfate days < 7.5 $\mu g/m^3$.

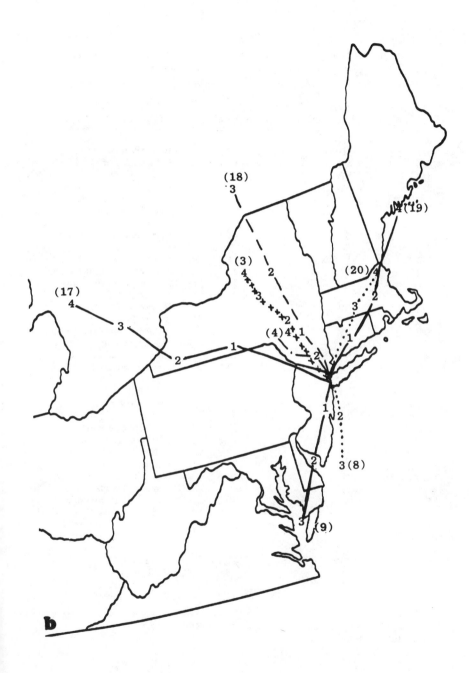

TABLE 1

COMPARISON OF WIND TRAJECTORY WITH ACTUAL SULFATE CONCENTRATIONS
FOR THE NEW YORK SUMMER AEROSOL STUDY

Date (Starting)	Average Trajectory Direction	$SO_4^=$ Actual ($\mu g/m^3$)	Weather Summary From Lippmann et al.[6]
8/2	NW	4.5 *	New air mass
8/3	NW–W	8.9	Westerly flow begins
8/4	SW	17.9	Return flow of H †
8/5	SW	25.4	Return flow of H
8/6	SW	26.2	Return flow—cold front approaching
8/7	SW–S	11.0	Cold front stagnates over NYC
8/8	S	4.7	Storm develops in front
8/9	S–N	7.1	Hurricane Belle
8/10	W (variable)	14.3	Hurricane Belle
8/11	W	22.0	New air mass that had been sitting over the Midwest since Aug. 7
8/12	SW	36.9	Return flow of H
8/13	SW	36.7	Return flow of H
8/14	SW	20.9	Return flow—Cold front approaching
8/15	SW–NW	14.4	Cold front
8/16	NW–W	3.7	New air mass—brisk NW flow
8/17	N–NE	2.5	Brisk N flow
8/18	NE	3.3	Brisk NE flow
8/19	NE	3.6	NE flow
8/20	NE–W	7.4	Westerly flow starts
8/21	W	15.6	Return flow of H
8/22	W	32.0	Return flow of H
8/23	W–N	22.0	Return flow of H
8/24	N	11.6	New air mass as cold front moves through
8/25	W	40.2	New air mass assimilated into air mass of Aug. 23
8/26	SW	31.4	Return flow of same H
8/27	SW	24.1	Disorganized pattern of return flow
8/28	SW	22.1	Same as 8/27
8/29	N	5.0	Cold front
8/30	NW	2.1	New air mass
8/31	W	12.9	Westerly flow starting

* > 7.5 $\mu g/m^3$—high $SO_4^=$ day; < 7.5 $\mu g/m^3$—low $SO_4^=$ day.
† H=High pressure system.

Results of speciation and simultaneous size fractionation analyses are also compatible with the transport phenomenon. Tanner *et al.*[7] found most of the $SO_4^=$ to be in the fine particle size range. These types of particles can have lifetimes in the atmosphere on the order of days and would be transportable. The $SO_4^=$ made up 70% of the submicron aerosol and 50% of that in the range below 0.25 μm in diameter. The speciation analyses in the NYSAS also indicated that there was a high correlation between the ammonium (NH_4^+) and $SO_4^=$ found in the accumulation size range with the NH_4^+ accounting for 85% of the variation in the $SO_4^=$ concentration.[7] Knowing the above, and the relationship of the $SO_4^=$ with the air mass, the method for $SO_4^=$ production is consistent with the homogenous reaction mechanism started by the reaction of ozone (O_3) with an olefin and followed by a complex series of reactions with hydroxyl (OH) which produce hydroperoxyl (HO_2) and peroxyalkyl (RO_2) radicals. The reaction of SO_2 directly with OH forms HSO_3 at a rate several orders of magnitude faster than the others. The fate of HSO_3 is somewhat uncertain, but it appears to undergo a series of reactions with NO_2, O_2, and RH to form H_2SO_4. The reactions with HO_2 and RO_2 are straight forward and involve:

$$HO_2 + SO_2 \longrightarrow SO_3 + OH \tag{1}$$

$$RO_2 + SO_2 \longrightarrow SO_3 + RO \tag{2}$$

which are followed by:

$$SO_3 + H_2O \longrightarrow H_2SO_4 \tag{3}$$

Additional neutralizing reactions then occur with gaseous ammonia (NH_3):

$$H_2SO_4 \longrightarrow NH_4HSO_4 \tag{4}$$
$$\text{(ammonium bisulfate)}$$

and

$$NH_4HSO_4 + NH_3 \longrightarrow (NH)_2SO_4 \tag{5}$$
$$\text{(ammonium bisulfate)}$$

The primary or initiating species is ozone which is readily available in the summertime air masses investigated in this study. In light of this reaction scheme and the $SO_4^=$ being primarily in the form of NH_4^+ compounds, most of the aerosol detected in New York City was probably aged and caused by transformation and transport on a regional scale. In contrast, the strong acid species that accounted for the remaining variability in the sulfate were associated with smaller particles and were probably fresh aerosol attributable to local production.

SUMMARY AND CONCLUSIONS

An average of 35% of the TSP in New York was attributed to long-range transport, and this was assumed to be essentially all in the accumulation mode. Of this, up to about half of it is sulfate on the basis of the estimate that 73% of the sulfate is due to transport and that up to 25% of the New York TSP is sulfate, i.e., $0.73 \times .25 = .18$. The other half of the transported aerosol appears to include organics, trace metals, and crustal type materials. Most nitrates are probably of local origin because they are found primarily in particles >3.5 μm in diameter.[8] The results from the fine particle sampling in the diameter range from 0.1 μm to 0.5 μm appear to substantiate this since only a slight diurnal pattern was found in this range.[2] Sulfate accounts for as much as half of the accumulation mode mass concentration, but the contributions are weighted toward the larger particles within this mode. In contrast, particles <0.1 μm in diameter have a discernable diurnal pattern, which correlates to local traffic emissions.[2] Coagulation of these fine particles produced increases in accumulation-mode concentrations, but since this range has a weak diurnal pattern, the local contribution from growth was probably slight.

This discussion indicates that total suspended particulate and sulfate transport can affect the metropolitan area throughout the summer. Further substantiation is being attempted in a second New York Summer Aerosol Study, but the conclusions from existing data suggest that future control strategies for total suspended particulates and for chemical fractions found in the TSP and accumulation mode suspended particulates must account for the amounts being transported into the New York area from relatively distant regions.

REFERENCES

1. WOLFF, G. T., P. J. LIOY, B. P. LEADERER, D. M. BERNSTEIN & M. T. KLEINMAN. 1979. Characterization of Aerosols Upwind of New York City, I. Transport. Ann. N.Y. Acad. Sci. **322**. (This volume)
2. KNUTSON, E. O., D. SINCLAIR & B. P. LEADERER. 1979. New York Summer Aerosol—Number Concentration and Size Distribution of Atmospheric Particles. Ann. N.Y. Acad. Sci. **322**. (This volume)
3. LEADERER, B. P., D. ROMANO & J. A. J. STOLWIJK. 1979. Light Scattering Measurements of the New York Summer Aerosol. Ann. N.Y. Acad. Sci. **322**. (This volume)
4. HIDY, G. M. 1974. Characterization of Aerosols in California (ACHEX). Volumes 1, 2, 3, and 4. Prepared for California Air Resources Board. Thousand Oaks, CA. Rockwell International.
5. WHITBY, K. T., R. B. HUSAR & B. Y. H. LIU. 1972. The Aerosol Size Distribution of Los Angeles Smog. *In* Aerosols and Atmospheric Chemistry. G. M. Hidy, Ed. Academic Press. New York.
6. LIPPMANN, M., M. T. KLEINMAN, D. M. BERNSTEIN, G. T. WOLFF & B. P. LEADERER. 1979. Size-Mass Distribution of the N.Y. Summer Aerosol. Ann. N.Y. Acad. Sci. **322**. (This volume).
7. TANNER, R., R. GARBER, W. MARLOW, B. P. LEADERER & M. A. LEYKO. 1979. Chemical Composition and Size Distribution of Sulfate as a Function of Particle Size in the New York Aerosol. Ann. N.Y. Acad. Sci. **322**. (This volume).
8. KLEINMAN, M. T., C. TOMCZYK, R. TANNER & B. P. LEADERER. Inorganic Nitrogen Compounds in the New York City Air. Ann. N.Y. Acad. Sci. **322**. (This volume).
9. DAISEY, J. M., M. A. LEYKO & E. HOFFMAN. The Nature of the Organic Fraction of the New York Summer Aerosol. Ann. N.Y. Acad. Sci. **322**. (This volume).
10. AMES, B. N., J. McCANN & E. YAMASAKI. 1975. Methods for detecting carcinogens and mutagens with the Salmonella/mammalian microsome mutagenicity test. Mutation Res. **31:** 347–364.
11. BERNSTEIN, D. M. & K. RAHN. 1979. New York Summer Aerosol Study: Trace Element Concentration. Ann. N.Y. Acad. Sci. (This volume).
12. KNEIP, T. J., B. P. LEADERER, D. M. BERNSTEIN & G. T. WOLFF. 1979. The New York Summer Aerosol Study (NYSAS), 1976. Ann. N.Y. Acad. Sci. **322**. (This volume).

13. LIOY, P. J., G. T. WOLFF, K. A. RAHN, D. M. BERNSTEIN & M. T. KLEINMAN. 1979. Characterization of Aerosols Upwind of New York City, II. Chemical Composition. Ann. N.Y. Acad. Sci. **322**. (This volume).
14. WOLFF, G. T. & P. J. LIOY. 1977. Transport of Suspended Particulates into the New York Metropolitan Area. *In* Proceedings of the 70th Annual Meeting of the Air Pollution Control Association, No. 77.2.7.
15. WOLFF, G. T., P. J. LIOY, R. E. MEYERS, R. T. CEDERWALL, G. D. WIGHT, R. S. TAYLOR & R. E. PASCERI. 1977. Anatomy of Two Ozone Transport Episodes in the Washington, D.C. to Boston, Mass. Corridor. Environ. Sci Technol.
16. LEADERER, B. P. 1977. Personal communication.
17. KLEINMAN, M. T., T. J. KNEIP & M. EISENBUD. 1976. Seasonal Patterns of Airborne Particulate Concentrations in New York City. Atmos. Environ. **10**(1): 9–11.